Human Herpesvirus Infection - Biological Features, Transmission, Symptoms, Diagnosis and Treatment

Edited by Ronaldo Luis Thomasini

Published in London, United Kingdom

IntechOpen

Supporting open minds since 2005

Human Herpesvirus Infection - Biological Features, Transmission, Symptoms, Diagnosis and Treatment
http://dx.doi.org/10.5772/intechopen.73940
Edited by Ronaldo Luis Thomasini

Contributors
Şule Gökçe, Emmanuel Drouet, Vittorio Mazzarello, Marco Ferrari, Maria Alessandra Sotgiu, Stefano Decandia, Hideto Nakajima, Makoto Hara, Akihiko Morita, Satoshi Kamei, Ronaldo Luis Thomasini, Vesna Martic

First published in London, United Kingdom, 2020 by IntechOpen
IntechOpen is the global imprint of INTECHOPEN LIMITED, registered in England and Wales, registration number: 11086078, 7th floor, 10 Lower Thames Street, London, EC3R 6AF, United Kingdom
Printed in Croatia

British Library Cataloguing-in-Publication Data
A catalogue record for this book is available from the British Library

Additional hard and PDF copies can be obtained from orders@intechopen.com

Human Herpesvirus Infection - Biological Features, Transmission, Symptoms, Diagnosis and Treatment
Edited by Ronaldo Luis Thomasini
p. cm.
Print ISBN 978-1-83881-158-7
Online ISBN 978-1-83881-159-4
eBook (PDF) ISBN 978-1-83881-160-0

Meet the editor

Ronaldo Luis Thomasini is a Professor of Microbiology, Parasitology, Pathology, and Pharmacology at the Faculty of Medicine of Diamantina-Federal University of Jequitinhonha and Mucuri Valleys - UFVJM, Brazil. He is a Coordinator of The Center for Studies of Inflammatory and Infectious Diseases and a Professor in the Multi-Centric Post-Graduation Program in Physiological Sciences (PMPGCF-UFVJM), and Post-Graduation Program in Pharmaceutical Sciences (PPGCiFarm-UFVJM). He is a graduate (Licenciate and Bachelor) in Biological Sciences by Herminio Ometto University Center (2002), Master degree in Pharmacology by State University of Campinas - UNICAMP (2007), and PhD in Clinical Medicine by the same university (2011). He has a post-doctoral fellow in Biochemistry and Immunology from the Federal University of Minas Gerais-UFMG and develops studies in clinical pathology, microbiology, virology, immunology and biochemistry, techniques on molecular biology applied to the diagnosis of infectious disease and host-parasite interactions.

Contents

Preface

The relationship between herpesviruses and humans probably dates from thousands of years ago. In the last few decades, many aspects of herpesvirus infections have been understood, such as infections with a severe outcome to mild or subclinical manifestations. Herpesvirus can remain latent during a lifetime and sometimes their reactivation can cause different clinical features in the patient. Several conditions have been related to herpesvirus reactivation such as complications in transplant organ recipients and immune regulatory modification in the elderly. In this book, aspects of human herpesvirus simples, varicella-zoster virus, and cytomegalovirus have been presented and discussed by our eminent colleagues.

As the editor, I thank all of the authors for their contribution to this book and the sharing of their experience on herpesvirus infections with researchers, physicians, other professionals, and students around the world. I thank you for your hard work.

To my colleagues at the Federal University of Jequitinhonha and Mucuri Valleys and others whom, during my life, have contributed to scientific insights and/or friendship, you have molded my ethos and professional features. To my wife Fabiana and my children Helena and Miguel for inspirations of the soul and my lovely life. I love you all. To my parents Tereza and Osvaldo (*in memorial*) for education given at the dawn of my life and all other relatives and friends. Finally, I thank Mrs. Lada Bozic for her assistance in turning this publication in to a 'real dream'.

Thank you all.

Ronaldo Luis Thomasini PhD, MSc, BSc.
Professor,
Faculty of Medicine of Diamantina,
Federal University of Jequitinhonha and Mucuri Valleys,
Brazil

Section 1

Introduction

Chapter 1

Introductory Chapter: Human Herpesvirus - A Short Introduction

Ronaldo Luis Thomasini

1. Introduction

The relationship between herpesviruses and humans probably proceeds from thousands of years. In the last decades, many aspects of herpesviral infections have been understood since infections with a severe outcome to mild or subclinical manifestations. Several conditions have been related to herpesviral reactivation since complications in transplant-organ recipients to immune regulatory modification in the elderly.

The manifestations caused by the herpesvirus simples in the oral region are classically known. Manifestations with labial lesions and sometimes stomatitis can occur in a considerable part of the population at some point in the individual's life. Usually, these lip lesions are self-limiting although they are often recurrent [1]. Other herpesviruses have been considered as emerging pathogens in the etiology of different diseases.

Human herpesviruses belong to the family *Herpesviridae*, they are ubiquitous viruses and once the first infection occurs, they remain in the body of the affected individual (latency) during the lifetime. These viruses cause a wide variety of diseases, and infections are often benign, but may in immunocompromised individuals cause clinical manifestations with different level of severity [2, 3].

The *Herpesviridae* family is divided into 3 subfamilies: Alphaherpesvirinae (*α-herpesvirinae*), Betaherpesvirinae (*β-herpesvirinae*) and Gammaherpesvirinae (*γ-herpesvirinae*). These are distinguished by their viral and structural characteristics, as well as by their pathogenic potential. All types of viruses classified into this

Virus	Synonymous	Subfamily	Abbreviation
Human herpesvirus 1	Herpes simplex-1	α	HSV-1/HHV-1
Human herpesvirus 2	Herpes simplex-2	α	HSV-2/HHV-2
Human herpesvirus 3	Varicella-zoster	α	VZV/HHV-3
Human herpesvirus 4	Epstein-Barr	γ	EBV/HSV-4
Human herpesvirus 5	Cytomegalovirus	β	CMV/HHV-5
Human herpesvirus 6	None	β	HHV-6
Human herpesvirus 7	None	β	HHV-7
Human herpesvirus 8	None	γ	KSHV/HHV-8

Table 1.
Members of the human herpesvirus family.

IntechOpen

family are double-stranded DNA viruses and the different types of herpes viruses share similar structural characteristics. List of herpesviruses that infect humans are summarized in **Table 1** [2, 3].

Herpesvirus group can establish primary infections with nonsevere symptoms, which can result in an efficient immune response that prevents a new infection. However, the virus is not completely eliminated, its genome remains within cells without productive infection. Latent infections may become active (reactivation) due to factors related to the host and these manifestations allow the spread of herpesviruses, since a release of extracellular virions occurs which can infect other cells [4].

2. The human herpesviruses

2.1 Herpesvirus simplex 1 and 2

Human herpesviruses type 1 and 2 (HSV-1 and HSV-2) are usually associated with herpes and genital herpes, respectively. However, genital herpes may be a consequence of HSV-1 infection and cold sores may also be caused by HSV-2. Once the individual has been infected, reactivation is extremely common in both clinical forms: oral or genital. The lesions are bullous and painful although it tends to disappear in a few intervals of time [5, 6].

In some individual, especially in severely immunocompromised patients but not only but also in individuals with a moderate or mild reduction in immune response, in particular, cell-mediated, these viruses can cause more severe disease such central nervous system affections. The frequent labial and genital herpes recurrence show that severe immunosuppression is not a *sine qua non* condition to herpes simples reactivation.

2.2 Varicella-zoster virus

Human herpesvirus type 3 (Varicella-zoster) causes varicella (chickenpox) in a primary infection that occurs especially in children and reactivation can cause the onset of zoster that is more frequent in the elderly. Neurologic zoster is an important complication of this infection, especially in elderly and immunosuppressed patients [7, 8].

Varicella in children is often benign although rarely could cause hepatitis and encephalitis. In adults, varicella tends to cause more severe outcomes, and the fact is not completely understood. Zoster is related to decrease of the self-reported level of satisfaction with the life and depression in the elderly. Besides the zoster being frequently recurrent, the pain remains for several weeks after the lesions have disappeared. Fortunately, the vaccination is available for both children [9] and elderly [10] and it will probably lead to a decrease of incidence not only of varicella but also of zoster due to the consequence of reactivation of VZV which remained from a Varicella episode. However, in the elderly, the prescription of vaccination must be carefully made to avoid complications caused by a vaccine containing attenuated viruses.

2.3 Epstein-Barr virus

Human herpesvirus type 4 (Epstein-Barr virus; EBV) is associated with infectious mononucleosis, Burkitt's lymphoma, and nasopharyngeal carcinoma. The most important aspect of EBV is its oncogenic potential. A majority of cases of

EBV infection have a low impact on the individual; however, a complication such as hepatitis [11], lymphoproliferative syndrome [12], and encephalitis can rarely occur [13]. There is no vaccine to prevent EBV infection; however, concerning a progressive putative relation with the pathogenesis of several types of tumors, a possibility of the development of a vaccine must be considered.

2.4 Cytomegalovirus

Primary cytomegalovirus infection causes a "mononucleosis-like syndrome" known as cytomegaly or "cytomegalic inclusion disease." Cytomegalovirus is not well known by the general public because it causes often not severe clinical conditions. The impact of cytomegalovirus infection HIV-infected [14] and transplant patients [15] is well recognized. Most recently, the cytomegalovirus has been associated with immunosenescence and, perhaps, frailty syndrome in the elderly [16]. It is not clear whether is the virus that modulates the immune system as a mechanism of evasion or the immune system declines along with the aging and to allow the virus replication.

2.5 Human herpesviruses 6 and 7

Primary HHV-6 and HHV-7 infections cause a common early febrile infectious syndrome known as roseola infantum or exanthem subitum [17]. The HHV-6 has been related to transplant rejection and graft-versus-host disease in bone marrow transplantation [18]. In other types of transplantation, the HHV-6 effects are much less related. HHV-7 have been studied in several types of conditions but its influence remains still not clear.

2.6 Human herpesviruses 8

Human herpesvirus type 8 is associated with Kaposi's sarcoma and can lead to death in immunosuppressed patients, especially acquired immunodeficiency syndrome (HIV/AIDS) [19].

3. Conclusions

In the last decades, a large number of research regarding herpesviruses have been published. Although a considerable number of conditions are irrefutably linked to a specific type of herpesvirus, many others remain well less explained and with lack of cause-effect definitions. It is clear that further studies must be designed with the aim of better understanding each hypothesis. The *status quo* about the knowledge about the human herpesvirus simples, varicella-zoster virus, and cytomegalovirus was presented and discussed in this book.

Conflict of interest

The author declares that there is no conflict of interest.

Author details

Ronaldo Luis Thomasini
Faculty of Medicine of Diamantina, Federal University of Jequitinhonha and Mucuri Valleys, Diamantina, Brazil

*Address all correspondence to: ronaldothomasini@gmail.com

References

[1] Bannoehr J, Franco A, Iurescia M, Battisti A, Fitzgerald JR. Koneman. Koneman's color atlas and textbook of diagnostic microbiology. Journal of Clinical Microbiology. 2009;**47**: 469-471

[2] Grinde B. Herpesviruses: Latency and reactivation - viral strategies and host response. Journal of Oral Microbiology. 2013;**5**:22766

[3] Fishman JA. Overview: Cytomegalovirus and the herpesviruses in transplantation. American Journal of Transplantation. 2013;**13**(s3):1-8

[4] Cohrs RJ, Gilden DH. Human herpesvirus latency. Brain Pathology. 2001;**11**(4):465-474

[5] Whitley RJ. Herpesviruses. In: Medical Microbiology. Galveston, Texas, USA: University of Texas Medical Branch at Galveston; 1996

[6] Wilck MB, Zuckerman RA, Zuckerman RA. Herpes simplex virus in solid organ transplantation. American Journal of Transplantation. 2013;**13**:121-127

[7] Pergam SA, Limaye AP. Varicella zoster virus in solid organ transplantation: Guidelines from the American Society of Transplantation infectious diseases Community of Practice. Clinical Transplantation. 2019;**4**:e13622

[8] Schmader K. Herpes zoster in the elderly: Issues related to geriatrics. Clinical Infectious Diseases. 1999;**28**(4):736-739

[9] World Health Organization. Varicella and herpes zoster vaccines: WHO position paper, June 2014 – Recommendations. Vaccine. 2016;**34**(2):198-199

[10] Schmader KE, Johnson GR, Saddier P, Ciarleglio M, Wang WWB, Zhang JH, et al. Effect of a zoster vaccine on herpes zoster-related interference with functional status and health-related quality-of-life measures in older adults. Journal of the American Geriatrics Society. 2010;**58**(9):1634-1641

[11] Moniri A, Tabarsi P, Marjani M, Doosti Z. Acute Epstein-Barr virus hepatitis without mononucleosis syndrome: A case report. Gastroenterol Hepatol from Bed to Bench. 2017;**10**(2):147-149

[12] Kim H-J, Ko YH, Kim JE, Lee S-S, Lee H, Park G, et al. Epstein-Barr virus-associated lymphoproliferative disorders: Review and update on 2016 WHO classification. Journal of Pathology and Translational Medicine. 2017;**51**(4):352-358

[13] American Academy of Neurology. Singh DT, Rabinstein A. Neurology [Internet]. Vol. 86, Neurology. Advanstar Communications; 2016. P1.293

[14] Gianella S, Letendre S. Cytomegalovirus and HIV: A dangerous pas de Deux. The Journal of Infectious Diseases. 2016;**214** (Suppl 2):S67-S74

[15] Azevedo LS, Pierrotti LC, Abdala E, Costa SF, Strabelli TMV, Campos SV, et al. Cytomegalovirus infection in transplant recipients. Clinics (São Paulo, Brazil). 2015;**70**(7):515-523

[16] Thomasini RL, Pereira DS, Pereira FSM, Mateo EC, Mota TN, Guimarães GG, et al. Aged-associated cytomegalovirus and Epstein-Barr virus reactivation and cytomegalovirus relationship with the frailty syndrome in older women. PLoS One. 2017;**12**(7):e0180841

[17] Wolz MM, Sciallis GF, Pittelkow MR. Human herpesviruses 6, 7, and 8 from a dermatologic perspective. Mayo Clinic Proceedings. 2012;**87**(10):1004-1014

[18] Pichereau C, Desseaux K, Janin A, Scieux C, Peffault de Latour R, Xhaard A, et al. The complex relationship between human herpesvirus 6 and acute graft-versus-host disease. Biology of Blood and Marrow Transplantation. 2012;**18**(1):141-144

[19] Batista MD, Ferreira S, Sauer MM, Tomiyama H, Giret MTM, Pannuti CS, et al. High human herpesvirus 8 (HHV-8) prevalence, clinical correlates and high incidence among recently HIV-1-infected subjects in Sao Paulo, Brazil. Broliden K, organizador. PLoS One. 2009;**4**(5):e5613

Section 2

Human Herpesvirus Simples

Chapter 2

Sunlight and Herpes Virus

Vittorio Mazzarello, Marco Ferrari, Stefano Decandia and Maria Alessandra Sotgiu

Abstract

The Herpesviridae are a family of viruses widely spread in nature that can infect a wide variety of species. After the primary infection, the human alphaherpesvirinae sub-family remains quiescent in the nerve ganglia from which it can periodically reactivate, causing clinical manifestations. Although spontaneous recurrences are possible, a wide variety of internal and external triggers may lead to transformation of the Herpes Simplex and Varicella-Zoster Viruses from a dormant to a proliferative state. Sunlight is a potent stimulus for the alphaherpesvirinae reactivation. The purpose of this paper is to analyze various features of this correlation and several steps you can take to lower your risk of triggering a herpes outbreak after sun exposure. Learning how to reduce the recurrence is extremely important and it is necessary: to perform a gradual and progressive sun exposure; to know what garments to wear; to know the environmental conditions of exposure; to know each skin phototype; to use a protective product against UVB and UVA with sun protection factor suitable for each phototype and environmental conditions.

Keywords: sunlight, UV, UVR, herpes virus, HSV, herpes zoster, VZV, sunscreens, prevention

1. Introduction

The sunlight and specifically the Ultraviolet component of its radiation (UVR) is among the major causes of alphaherpesviridae (αHV) reactivation. Various aspects of this correlation will be analyzed in this chapter, as well as how it interferes with the virus-host relationship and what kind of precautions should be taken to reduce the risks of painful relapse.

2. Herpes virus

The Herpes Virus (HV) are a members of the family Herpesviridae widely spread in nature that can infect a wide variety of species of at least two animal phyla, the Chordata and the Mollusca [1]. It is a virus about 150–200 nm in diameter, with icosahedral nucleocapsid DNA double helix containing an envelope which derives from the nuclear membrane of the host cell with viral glycoproteins that protrude on the surface.

To date a total of 8 human HVs are known, having the characteristic of establishing a life-long latent infection: a state from which the virus can be reactivated and result in recurring disease. The HV family is divided into three subfamilies (Alphaherpesvirinae,

Betaherpesvirinae, and Gammaherpesvirinae); among these, only the αHV creates skin lesions in humans [2]. The Herpes Simplex Virus (HSV), creating the general clinical picture of herpetic disease, and the Varicella-Zoster Virus (VZV), which is the cause of chickenpox and Herpes Zoster (HZ), both belong to αHV.

2.1 Clinical aspects of αHV lesions

The transmission of αHV occurs by close contact with a person who actively eliminates the virus. The viral diffusion occurs from lesions; however, it can occur even if they are not visible. After the primary infection, the αHV remains quiescent in the nerve ganglia from which it can periodically reactivate, causing clinical manifestations. The HSV commonly cause a relapsing mucocutaneous infection affecting the skin, mouth, lips, eyes and genitals. Serious common variants include encephalitis, meningitis, neonatal herpes, and infections disseminated in immunosuppressed patients. There are two types of HSV: Herpes Simplex Virus 1 (HSV-1) and Herpes Simplex Virus 2 (HSV-2) and both types can cause oral or genital infections. In most cases, the HSV-1 causes gingivostomatitis, cold sores, herpetic keratitis, and lesions in the upper body. The HSV-2 generally causes lesions to the genitals and to the skin of the lower half of the body. Approximately 70% of population in USA is seropositive for HSV [3] but only the 20%, due to a decline in cellular immunity, presents the recurrent form that can occur with a variable frequency. The mucocutaneous manifestations occur in two forms: primary infection and recurrent infection. Both forms appear on the skin with an erythematous lesion with vesicles (the size of a pin's head) clustered that can merge to form a bubble and then break, leaving an erosion and then a crust that falls after a few days. The primary infection may be unapparent, so that most individuals carry antibodies but have no memory of the initial Herpes. In the forms where the disease manifests itself usually appears in children aged between 6 months and 3 years of age. It presents a clinical presentation often more serious than the classical that is shown in the recurrent form. In fact, it is associated with general malaise with temperature over 39°C, pain, dysphagia, sialorrhea, fetid breath. Despite the impressive appearance it resolves on its own in 10–15 days. During the primary infection the transmission of the virus is favored by alterations of the epithelial lining, so that it penetrates and multiplies in the epithelial cells, with lysis of the infected cells due to the formation of a large number of virions. The virus then disappears from the coating epithelium and goes, passing through the sensory nevi, to localize in the nerve ganglia corresponding to the entry area. In recurrent manifestations usually, after a prodromal period (typically <6 h in HSV-1 relapses) characterized by burning or pruritus, small vesicle bunches appear stretched on an erythematous base. The bunches are 0.5–1.5 cm in size but can flow together. Skin lesions of the nose, ears, eyes, fingers or genitals can be particularly painful. The vesicles normally persist for a few days, then break and dry, forming a thin yellowish crust. The lesions can be associated with a burning sensation, tingling or itching with or without fever and small adenopathies, the evolution lasts 1–2 weeks. The herpetic lesions typically heal completely, but recurrent lesions in the same site can cause atrophy and scarring. Skin lesions can develop bacterial superinfections. In patients with depression of cell-mediated immunity due to Human Immunodeficiency Virus infection (HIV) or other causes, long-lasting or progressive lesions may persist for weeks or longer. Herpes labialis (HL) occurs on the edge of the vermilion of the lip or, less frequently, on the mucosa of the hard palate. HL is the most common clinical form in the facial region [4]. In the United Kingdom it accounts for 1% of medical consultations [5]. The acute gingivostomatitis is characteristic of childhood. Instead, herpetic pharyngitides can occur in adults and children; occasionally, mediated by oro-genital contacts, caused by the HSV-2. The intraoral and gingival vesicles normally break in a time ranging from a few hours to 1–2 days, leaving an ulcer. Fever and pain often occur; after the resolution, the virus remains quiescent in the semilunar ganglion. The Genital Herpes (HG) is the most widespread sexually transmitted ulcerative disease in developed countries. HG can be caused by HSV-1 or HSV-2. Ocular HSV lesions (HO) can cause corneal scarring, and recurrent ocular HSV infections are a leading cause of vision loss [6]. VZV belongs to the αHV subfamily and produces two clinical syndromes: varicella (chickenpox) and zoster (shingles). Both "zoster" (from Greek) and "shingles" (from French and Latin languages) correspond to the English word "belt," which describes the characteristic narrow, bandlike rash from the spine to the front of the torso on one side of the body [7]. VZV is transmitted by inhalation of respiratory secretions or contact with skin lesions. During the primary infection (varicella), the virus becomes latent in the dorsal ganglia and zoster is due to reactivation from latency, a process which occurs most frequently in elderly [8]. Each person with a history of varicella has approximately a 30% lifetime risk of at least one VZV reactivation [9].

2.2 Pathogenesis

HSV infections are most commonly acquired through direct contact with mucosal tissue or secretions of another infected person and the majority of infections are established within the stratified squamous epithelium of the skin and oral or genital mucosa [10]. The virus is able to cause a lytic infection with direct death of epithelial cells. Following infection, the virus enters sensory nerves that innervate the skin or mucosa and travels via retrograde axonal transport to the neuronal cell body: here it can establish a life-long latent infection in dorsal root ganglia [11]. In the cell body there is the nucleus, where the virus makes use of the cell's apparatus for DNA replication and transcription. The axonal cytoskeleton and molecular motors, like kinesins, are involved in the active transport of viral capsids and glycoproteins: their transport seems to be fast, bidirectional and microtubules dependent [12–14]. The mechanisms that regulate entry into lytic replication versus latent infection in neurons remain largely undefined. The mechanism of HSV entry is mediated by direct interaction between viral envelope glycoproteins and cell surface receptors that mediate attachment, initiate signaling cascades, or trigger virus internalization.

The entry process involves multiple steps:

1. Attachment to the cell surface: the virus initially uses filopodial interaction to migrate toward the cell body and to initiate the access. This process is termed 'viral surfing' [15]. In this process, the initial binding of virus to cells is mediated through association of viral glycoprotein (g)B and/or gC with heparan sulfate proteoglycans (HSPG)s located on the cell surface, facilitating the subsequent binding to coreceptors. The gC makes the first contact with HSPGs on the cell surface, but in the absence of gC, gB can take over this function [16].

2. Binding to cell receptors and coreceptors: the major virus attachment is glycoprotein gD and the most studied coreceptor are nectin-1, herpes virus entry mediator or 3-O-sulfated HS. The interaction of the viral glycoproteins with these cell receptors induces conformational changes recruiting gB, gH, and gL for fusion of the viral envelope with the cell plasma membrane leading to viral penetration and capsid release in the cytoplasm [17]. The link between gD and its receptor can activate the gH/gL complex.

3. Fusion with cellular membrane: gH/gL provides the signal required for activation of gB. Binding of gB to one of its receptors, is required for

delivery of the viral nucleocapsid to the cytoplasm accomplished either by membrane fusion or endocytosis/phagocytosis-like uptake. Beside membrane fusion, mechanisms of endocytosis and/or a phagocytosis-like uptake have been proposed. The endocytosis of HSV particles is atypical, because not mediated by clathrin-coated pits or caveolae. The phagocytosis process requires a cytoskeletal rearrangement with activation of Rho GTPases [18].

After fusion between the cellular membrane with the infecting virus, a viral transactivator tegument protein (VP16), is released into the cytoplasm. The viral capsid is then transported to the nuclear membrane along the microtubule network and, through nuclear pore, the viral DNA is released into the nucleus. VP16 forms a transactivation complex binding in the cytoplasm host cell factor-1 (HCF-1) (protein that contains a nuclear localization sequence), and in the nucleus the homeodomain protein Octamer binding protein-1 (Oct-1). These proteins form a trimeric complex able to activate the immediate early (IE) gene expression [19]. Successful lytic replication is dependent on the expression of the viral IE genes within all infected cells. While this model of VP16 activation of IE gene expression is well understood, the mechanisms implicated in neuronal latency are debated and considerable gaps remain in our knowledge of how different signaling pathways act on the latent genome for reactivation. Following the establishment of latent infection, viral lytic gene expression is silenced, and the lytic gene promoters are associated with repressive heterochromatin [20]. Key experiments performed in the 1980s indicated that latent genomes in the brain stems of infected mice have a nucleosomal structure [21]. Later studies confirmed that the latent viral genome associates with cellular histones in the trigeminal ganglia of mice [22, 23]. Coinciding with the silencing of lytic transcripts, the viral lytic gene promoters become enriched with characteristic heterochromatic histone modifications [24, 25]. While it appears that factors intrinsic to neurons play a key role in the transcriptional silencing of the virus, viral gene products expressed during latent infection can also modulate the chromatin structure [23, 26, 27]. This modulation likely promotes long-term latency, while priming the genome for reactivation following the appropriate stimuli [28, 29].

3. Trigger

Although spontaneous recurrences are possible, a wide variety of internal and external triggers may lead to transformation of the HSV from a dormant to a proliferative state [30].

Some of the following factors may trigger herpes symptoms:

- Sunlight: some study demonstrates UVR as a powerful trigger for HL and it also seems that HZ can be stimulated by sun exposure [8, 31].

- Exposure to heat or cold [5, 32].

- Local tissue trauma may make herpes symptoms appear such as: undergoing a surgery [33], laser surgery [34], dental procedures [35], and Tattoos [36]. Another unusual form of traumatic triggering of HSV reactivation may be neurosurgery: after a delay of approximately 1 week, destructive encephalitis may develop with fever and seizures, and with typical viral inclusion bodies demonstrated by histopathology [37].

- Persistent mental stress and fatigue [33, 38]. Psychological stress can also dysregulate cellular immunity, and enhance latent αHV reactivation [39]. Importantly, chronically stressed low socioeconomic status individuals have higher antibody titers to latent HV. Additionally, dementia caregivers have greater HSV-1 antibody titers compared with demographically matched controls [40, 41].

- Physical stress: fever[5], illness (infection, septicaemia).

- Nerve damage: minimal stimulation or inapparent trauma to the trigeminal sensory root is sufficient to activate latent HSV in humans [42].

- Radiotherapy: the example of radiation therapy against a brain tumor initiating HSV encephalitis suggests that other trigger factors also should be studied [43].

- Immunosuppression: when the immune system is dysregulated, by HIV or chemotherapy or corticosteroid administration, people generally exhibit greater disease susceptibility and latent HSV or VZV reactivation [43, 44]. Maladaptive alterations in cellular immune function can enhance herpesvirus reactivation and replication, resulting in elevated herpesvirus antibody titers. For instance, organ transplant patients have elevated herpesvirus antibody titers [39, 45–47].

- Sexual intercourse: some people find that the friction of sexual intercourse irritates the skin and brings on symptoms of HG. Even if the friction of intercourse seems to be a trigger for symptoms, it won't probably cause a flare-up every time [5].

- Hormonal changes, like those that occur in the menstrual cycle, can affect herpes outbreaks. There is a significant association of development of recurrence HSV and the luteal phase of the menstrual cycle [5].

- Change in antiviral activity of the saliva [5].

Whether a common pathway exists for pathogenetic processes induced by these disparate reactivating factors remains to be determined.

4. The Sun radiation and interaction whit skin

The skin is continually subjected to the action of external agents including solar radiation. One of the scientifically documented triggers for herpes outbreaks is the ultraviolet (UV) light found in direct sunlight.

4.1 The Sun as origin of the electromagnetic energy

The Sun is a G-type main-sequence star and is the largest and the most massive object in the solar system. The Sun is the source of the overwhelming majority of light, heat, and energy on Earth's surface, and is powered by nuclear fusion of hydrogen nuclei into helium. As a result of these nuclear reactions a continuous flow of particles and electromagnetic waves called the solar wind is released in the cosmos. Solar wind is a constant stream of plasma and particles emanating from the sun and is the extension of solar corona into interplanetary space. The solar wind invests all the planets, can reach speeds above 700 km/s and have a density that varies from 10 to 100 particles/cm^3. Sunlight consists mostly of short wavelength ionizing radiation (cosmic, gamma, and X-rays) and long wavelength non-ionizing radiation (UV, visible, and infrared) [48].

UVR is the area of the electromagnetic spectrum that is considered biologically the most active and therefore of greatest impact on health and disease [49]. For convenience, we separate UV somewhat arbitrarily into UVA (315–400 nm), UVB (280–315 nm) and UVC (100–280 nm). UVC together with ionizing radiation is largely absorbed by the upper atmosphere and does not reach us on the earth's surface. Most UVR that reaches the earth's surface is UVA (95%), only a small percentage is UVB (approximately 5%). UVR peaks around noon and is increased by reflection from snow, water, and sand [50]. UVA, but not UVB, can penetrate glass [51]. The solar radiation is omnipresent during daylight hours. At ground level the amount of UV mainly comprises UVA, and a small percentage (<10%, variable by time of day, season and altitude) of UVB. The doses of UV absorbed vary greatly within a person and between people, depending on the position, time of day, season, type of clothing, habits and skin pigmentation.

The non-ionizing radiation are not lethal to living organisms but can cause damage to the skin and eyes if taken chronically and/or in large quantities. Animals defend themselves from the action of these waves thanks to the presence on their skin of hairs, feathers and scales. Humans, having lost the hair during evolution, have to use melanin as a means of protection. The peculiarity of the UV is that they are one of the few environmental factors that can cause both disease and protection against the disease [52]. The sun exposure is pleasant for us because it causes the following positive effects: we are pervaded by a pleasant feeling of warmth and well-being linked to Infrared Radiation (IR) and Visible Light (VL), we release chemical factors that act as antidepressants (VL), appears after a few hours a dark and transient tanning (UVA), followed by a golden and lasting tan (UVB) after 24–48 h. Other positive actions are the production of "antirachitic" vitamin D (UVB) and a regulation of hormonal functions (VL). Unsuitable exposure can lead to immediate or delayed side effects. The most frequent damages caused by sunlight are: sunburn, photoallergic reactions, photo-aging, skin tumors, eye diseases and immunosuppression.

4.2 UVR and immune skin suppression

Exposure to UVR has a profound effect on the skin immune system. It has both, pro-inflammatory as well as immunosuppressive effects and it involves both innate and adaptive immunity. Examples of pro-inflammatory responses clinically observed include sunburn, photodermatosis [53]. Examples of the immunosuppressive effect is the use of UV for psoriasis or lichen planus treatment. Both UVB and UVA wavebands contribute to sunlight-induced immunosuppression, although an interaction between them makes sunlight more suppressive than each waveband alone. It is therefore important to protect the skin from both UVB and UVA. Exposure to doses of UVR that are only 30–50% as high as what is required to cause barely detectable sunburn, suppressing immunity in humans. Therefore, normal daily outdoor activities during spring and summer months are likely to cause some degree of immunosuppression in a large proportion of humans [54]. It is both obvious and striking that UVR at rather low doses suppresses an immune response. Thus, one may speculate that a certain degree of immunosuppression may be beneficial. The skin is an organ which is constantly exposed to potential allergens; in addition, the skin is an organ which is prone to autoimmunity [55, 56]. Hence, it is tempting to speculate that a certain degree of constant immunosuppression by daily solar exposure may prevent the induction of these immune responses. Owing to the multiple different experimental systems suppressed by UV and the dependence on dose, timing, waveband and skin site, we currently do not have a comprehensive understanding of how UV has this potent effect on the immune system. However, many different molecular and cellular events have been described. The cells involved in immunosuppressive activity are keratinocytes, lymphocytes, Langerhans cells (LC), macrophages and mast cells. UVR induced immune suppression is known be mediated through T cells [57]. The relation of immune suppression is linked to various subtypes of regulatory immune cells such as regulatory T cells (Tregs) and regulatory B cells (Bregs) depends on UVR doses and type of immune response [58–61]. Furthermore, UVR has also profound effects on antigen-presenting cells. It damages LCs, so that they migrate from epidermis into the draining lymph nodes [62, 63]. It affects mast cells which are known to be involved in immune suppression [64]. It releases cytokines leading suppressor macrophages to infiltrate the skin and activating B lymphocytes in draining lymph nodes so that they have suppressor function. It is likely that interaction between these UV-altered antigen-presenting cells result in the activation of suppressor T lymphocytes. There is good evidence that these T suppressor cells are mainly responsible for reduction in immunity caused by UV [54]. The molecular mechanisms responsible for disruption of cellular immunity and some of the key events observed in the skin after the UVR exposure are described below (**Figure 1**). The cellular-molecular phenomena occur in successive steps. In the first step, which concerns keratinocytes, LC, urocanic acid (UCA) and corneum lipids, some ray-sensitive photoreceptors absorb photons, with different susceptibility for the different wavelengths (so the results can be different depending on the type of UV) and initiate a molecular cascade that damages and modifies the cellular biochemistry. The molecular mechanisms responsible for disruption of cellular immunity begins with DNA damage, trans to cis isomerization of UCA, and peroxidation of lipids. In the second step, the cells damaged by UVR produce mediators (especially cytokines) that modify the activity of LC. In fact, both for the cytokines and for their own damaged DNA, in addition to the alteration of the antigen presentation, they migrate into the lymph nodes. The cytokines produced in this phase are numerous. It has also been observed that UVR suppresses HSV antigen presentation in epidermal cells and leads to the reduction of type 1 cytokine release, an important key-factor in immunological control for viruses such as HSV [65, 66]. Photoproducts of DNA such as pyrimidine dimers or 6-4-photoproducts result in the production and release of various immunosuppressive factors such as Tumor Necrosis Factor (TNF)-alpha and

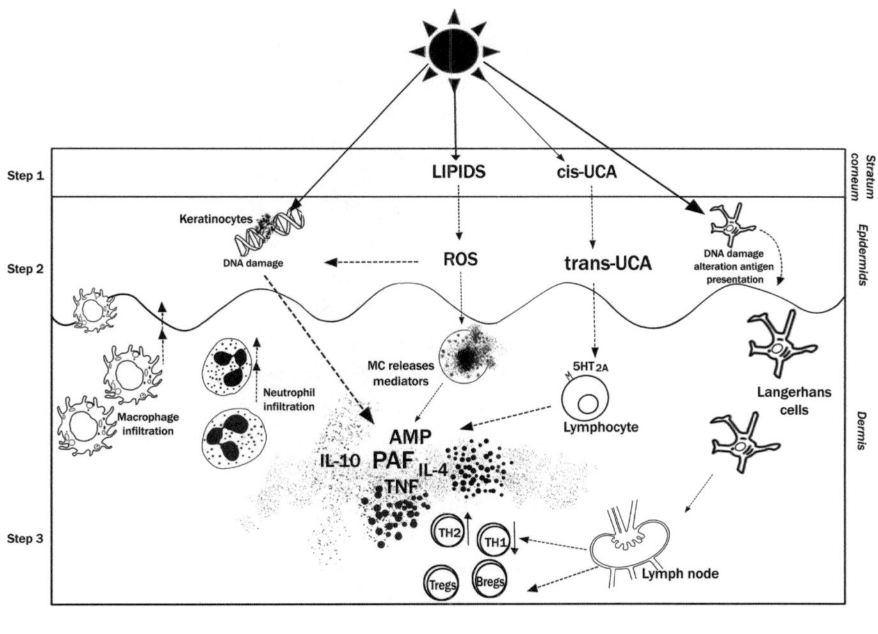

Figure 1.
The molecular mechanisms responsible for disruption of cellular immunity and some of the key events observed in the skin after the ultraviolet radiation exposure.

interleukin (IL)-10 by keratinocytes and other cells in the skin. The UVB waveband in particular also directly leads to isomerization of trans-UCA to cis-UCA. Cis-UCA induces immune suppression by binding to the 5-HT2A receptor, leading in turn to production of IL-10 by T-cells and B-cells. It may also indirectly lead to mast cell degranulation and stimulate the release of Platelet-Activating Factor (PAF). Formation of reactive oxygen species (ROS) by UVR not only induces and contributes to DNA damage but also directly stimulates PAF synthesis or the production of PAF-like molecules. UVR can also directly upregulate specific antimicrobial peptides (AMP) such as human beta-defensin-2, beta-defensin-3, S100A7, and RNase7 which are expressed by keratinocytes, lymphocytes, monocytes, and mast cells. These AMPs not only serve as initiators of innate immune response but they also communicate with the adaptive immune system and can activate it. The third step, as a result of the impact of UVR on the skin, is the appearance of an immunosuppressive microenvironment with abundance of TNF, IL-4 and IL-10 linked to Langerhans cell (LC) migration into lymph nodes and neutrophil and macrophage recruitment to the skin. As overall result, there is a modulation on T lymphocytes characterized by a global suppression of them and by a switch in the balance between two lymphocytes classes: the suppression of the Th1 population (implicated in immunity to intracellular organisms like viruses, through IL-2 and INF); an increase of Th2 (implicated in immunity against extracellular microbes such as bacteria, through IL-4/10) and an induction of Tregs and Bregs leading ultimately to functional immune suppression [67, 68].

4.3 Is UVR a cause of αHV recurrence?

A systematic epidemiological review was carried out in 2008 identifying 9 diseases that show sufficient evidence of a causal relationship with UVR exposure. These include the reactivation of the HSV. The other diseases are: melanoma, squamous cell carcinoma of the skin, basal cell carcinoma, solar keratoses, sunburns, cataracts, pterygium, squamous cell carcinoma of the cornea and conjunctiva [52]. In medical scientific literature several works have been published demonstrating the recurrence of αHV after exposition to solar UVR (sUVR) or experimental UVR (eUVR) both on human [69–74] and on animal models; [75, 76] due to these reasons most dermatology manuals recommend using sunscreen to avoid HSV recurrence [77]. Several papers have shown a correlation between UV exposure and occurrence of HSV-1 [74, 78]. Approximately the 25–50% of HL are attributed, at least in part, to sUVR exposure. In one scientific article it was shown that the use of sunscreen alone versus placebo showed 95–100% suppression of HL recurrences in 2 crossover trials after application of 4 Minimal Erythema Doses (MED) of eUVR [70, 79]. To evaluate the role of exposure to sUVR in primary and recurrent HSV-1 infections, the self-reported cause of infection among diagnosed patients in Hyogo Prefecture, Japan, was investigated. Among 4295 infected patients, 3678 had HSV-1, and 2656 of those patients (72.2%) had a recurrent flare-up. Sun-induced HSV-1 flare-up was reported by 10.4% of the total study population. However, this increased to 19.7% among patients diagnosed in July and August, to 28% among patients younger than 30 years diagnosed in July and August, and to 40% among patients younger than 30 years diagnosed in July and August with a recurrent infection [32]. Although these studies did not analyze HG, data from another study show that HG recurrences also occur more easily after exposure to UV rays. For example, one study found that patients with HG—in this case, on the buttocks—were likely to experience recurrences shortly after being exposed to eURV. Another study on HO compared the reactivation with sUVR, detecting an increase in reactivation in more exposed subjects, actually even if data are unclear due to confounding factors that can be superimposed, such as in particular the stress that might act both directly determining reactivation and indirectly probably creating a greater need to expose to the sun [80]. As far as the VZV is concerned, one work reported a higher incidence of total HZ cases and cases of zoster in males during summer (from July to September) with a significant increase in May–June in patients studied in 1992–1998 in Ferrara in north-east Italy [81]. Another work shows the incidence of HZ peak for all subjects and for males it coincides with the maximum UVR months in summer. This association was not found for women, considered alone. It has not been explained why this difference should occur between men and women, but one possible explanation could be that older men tend to have more activities outside than women, such as gardening or walking, and therefore more exposure to sUVR. In addition to the increase in the incidence of zoster in summer, there was a significant increase over the same period in cases where lesions occurred on the face were compared to body sites normally covered [8]. In addition to considering the possible influence of the seasons on the incidence of αHV mucocutaneous lesions, some studies have succeeded in demonstrating a correlation with the UVR dose, geographical location, age and the body location. A dose of eUVR capable of triggering the recurrence of HL, is 4 MED, which corresponds to 80 min of sun exposure around 12 in July, at sea level taken by an individual with unprotected fair skin [70]. In some works, a slight latitudinal gradient of HL and a peak of prevalence in adulthood are demonstrated [82–84]. Another work highlighted the photolocalization of viral exanthema by observing a particular distribution of skin lesions, especially for VZV and HSV, between exposed areas and areas covered by clothes, preferring a location exposed to rays [85]. The recurrence time of HL after eVR exposure may be immediate (within 48 h) or delayed (after 2–7 days); [78] the time required for virus reactivation at the latency site [86], virus transport to the skin surface (it is estimated that the speed, demonstrated in vitro, is 3–5 mm/h)] and the virus replication in the epithelium with production of typical lesions (>24 h) [87, 88]. The eUVR had a beneficial effect on the virulence of HSV in an animal model. In fact, in a study it has been shown that 80% of mice irradiated before infection, and then re-irradiated several weeks later, developed recrudescent lesions. Only 20% of equivalent mice had not been irradiated before infection, but when irradiated after infection developed recrudescences [85].

4.4 How does solar radiation stimulate viral reactivation?

The exposure to sunlight has been associated with HSV reactivation [89–91]. It has been observed that 30% of causes of reactivation and axon migration to the skin are due to sudden exposure to sunlight and this seems also linked to the triggering of various mechanisms. There are many ways in which UV exposure is thought to impact αHV, and HSV recurrence in particular, directly through 3 pathways and probably also indirectly with unknown methods [85]. The first pathway is the depression of immune response due to UV exposure. The second pathway by which UVR may affect recurrence is directly through HSV reactivation [80, 85]. The third pathway study molecular events that trigger reactivation. The first pathway is based on the hypothesis that the virus continually tends to migrate from the ganglion to the skin. According to this theory, the normal immune response is activated through cell-mediated mechanisms of lymphocytes and macrophages and through the release of cytokines. In this way most of the migrations of ganglion-to-skin viruses is suppressed, as they are represented by few viral units and because the system is already sensitized, preventing a clinically evident reactivation because the infection remains sub-clinical. In the first pathway, the exposure to UVR determines the imbalance and suppression of the immune system, in a dose-dependent manner, which triggers a series of events so that local control of the reactivation is lost causing some virions to escape from immune control and the disease becomes manifest. It does not seem that through this mechanism we can identify a "remote" influence that reactivates the virus, but only a local effect of more peripheral virions approaching the skin. In the second pathway, UVR directly determine an imbalance or radiation damage to epidermal and dermal cells, which are stimulated to repair producing transcription factors that in addition to activating cellular gene expression also activate the viral one and also inhibit the stimulus to apoptosis [80]. Especially the cell repair, through the c-Jun and c-Fos transcription factors, activates the HSV transcription promoter (infected cell polypeptide 0), leading to HSV transcription and reactivation [92]. Additionally, these repair pathways circumvent the activity of HSV latency-associated transcript preventing infected neurons from undergoing apoptosis and in turn, reactivating HSV [93]. Despite these models, significant gaps remain in our understanding of how these stimuli correlate with reactivation of the virus resulting in clinical disease. The third pathway is a molecular model that explains how UVR at the body surface results in multiple neuronal effects or hormonal alteration that could be relevant to reactivation. For example, a damage to innervated tissues that results in loss of the neurotrophin-producing cells and changes in the levels of regulatory neuropeptides, neurotrophins, neurotransmitters may occur following UV irradiation [94]. Nerve growth factor (NGF) deprivation was first found to trigger HSV reactivation in primary neuronal models of HSV latency using rat sympathetic neurons [95]. In vivo injection of anti-NGF serum into latently infected rabbits has also been shown to enhance reactivation of HSV [96]. Furthermore, interruption of signals downstream of the NGF receptor triggered reactivation in a variety of in vitro models of HSV latency [97–100], and has been shown to enhance explant mediated reactivation ex vivo [101, 102]. In addition, UV treatment in mice results in increased serum levels of cortisol and may act through a pathway that is similar to psychological stress-induced reactivation. It was also noted that the dexamethasone, a synthetic corticosteroid, stimulates reactivation of HSV-1 both ex vivo and in primary neuronal cultures, and the closely related bovine HSV-1 can also be reactivated in latently infected calves by intravenous injection of dexamethasone [98, 101, 103].

4.5 Do sun-screen reduce HSV recurrence?

To date four studies have been published on sunscreen used by volunteers who suffered from HL, two studies in which subjects were exposed to eUVR and two to sUVR. Two randomized controlled trials with a crossover design demonstrated, using a solar simulator, the effectiveness of lip sunscreen in reducing HL after UV exposure. The first study was conducted on 38 patients: it showed that after exposure to artificial ultraviolet, equal to 4 MED, HL developed in 27 patients (71%) treated with placebo. In contrast, when a sun protection factor (SPF) 15 sunscreen was applied during UV exposure, no lesion developed on 35 patients [70]. The second work carried out on 19 individuals, exposed to 4 MED for 10 min of ultraviolet light under artificial conditions, found that sunscreen significantly reduces relapses compared to placebo: one on 19 patients (5%) with sun protection against 11 out of 19 individuals (58%) with placebo [79].

Studies carried out in the natural environment have given different results.

The first work has been carried out in natural conditions in three ski resorts: Park City, Utah (January 21–28) SnowMass, Colorado (February 25 to March 3) and Keystone, Colorado (April 8–15) at a latitude between 40 and 39°. Fifty-one volunteer skiers were analyzed, showing that a SPF 15 sun screen compared to placebo was not effective to prevent reactivation of the virus. HL developed in 3 out of 24 subjects using protection and in 3 out of 27 with placebo [104]. This work was criticized by stating that the UV dose received by volunteer skiers during the trial was 1–3 MED per day, which is lower than the 4 MED needed to trigger recurrence [70]. Probably due to this limitation, this study is not mentioned in the main guidelines for HL treatment [105]. Furthermore, it is not reported what amount of sunscreen was applied by skiers. However, in the latter two experiments carried out with artificial light the sunscreens were likely applied in a dose sufficient to respect the SPF value [106]. The second randomized, crossover study was carried out in northern Sardinia (Italy) on 20 volunteers who went to beach at a latitude of 40–41° using a sunblock stick with SPF 30. The study was conducted between May and July 2017 around the summer solstice (June 21st) when the sun reaches its highest point in the sky, to make the total amount of solar irradiance equal in the two sequential study periods. For each volunteer the study period lasted 60 days: 30 with protection and 30 without protection. The month with or without product application was randomly assigned to each patient so as 10 subjects started the trial without protection and 10 with protection and the opposite during the following month. During the month when volunteers had to use a protection, they were requested to apply the sunblock stick on the vermilion and lip skin two times consecutively creating a double protective layer before going out or going to the sea. The protection was repeated every 2 h, after eating or drinking, smoking and after a swim. 4 MED were reached and exceeded by volunteers several times during the 2 months of study. In fact, each volunteer remained at the beach at around 12 am with an average of 4.5 ± 0.95 h in the period with stick and 4.3 ± 0.94 h in the period without stick exceeding the aforementioned dose. Results demonstrated that sunscreen is effective in protecting the upper lip from reactivating the HL. In fact, only one volunteer out of 20 had a HL during the period of sunscreen use versus 10 out of 20 without sunscreen during the studied period. One volunteer from the second group reported two sequential HL. The single event during the period with labial photoprotection was unleashed in the last week, the 11 events of the period without photoprotection appeared from the second week of exposure. All lesions were clinically diagnosed with the help of Tzanck's cytodiagnostic examination [107]. In summary, these three studies, even though with a small number of subjects, showed that sunscreens can reduce the relapses caused by HSV following UVR exposure both in the laboratory and in the open air.

5. Treatment

If you suffer from relapsing HSV or you want to reduce the risk of the onset of HZ especially in summer, the most effective way is to avoid sunlight. Obviously, this is not always possible for most people. Even if someone deliberately avoids going to the beach, the face and other exposed parts of the body will still come in contact with direct sunlight throughout the day. What should be done to avoid solar radiation or minimize its effects?

5.1 Practical photoprotection strategy

To minimize the risk of a HS recurrence it is necessary: to perform a gradual and progressive sun exposure; to know what garments to wear; to know the environmental conditions of exposure; to know each skin phototype; to use a protective product against UVB and UVA with SPF suitable for each phototype and environmental conditions. Sun exposure must be gradual and progressive. The ideal would be a tanning obtained with irradiation times that do not induce erythema for long periods, in order to activate mechanisms of natural photoprotection. In fact, it has been shown that sub-erythematous doses of UVB produce a tan. It is advised wearing long trousers and long-sleeved shirts during summer to avoid exposing more skin than necessary to direct sunlight and also a hat to protect the face from direct sunlight and to prevent lips and face from coming into direct contact with UVR. However, a garment does not offer a complete UV barrier. If it is wet, it has less dry effect in stopping UV. Dark colors absorb more UVR, while clear colors are more effective against IR. Cotton has a low protective factor compared to silk and blue jeans. We must therefore choose thick and darker fabrics to have an effective protection such as blue jeans. The effects of solar radiation also vary according to environmental conditions. For example, the amount of UVR in the environment varies during the day (maximum value between 12 and 16), in the different months of the year (period with more irradiation June–July–August in the northern hemisphere), in relation to the altitude (the quantity increases by 6% every km of height) and at the latitude (greater quantity in the tropics). When the sky is uniformly covered there is a reduction of about 50% of the UVR compared to the clear sky, but if it is partially cloudy the irradiation is not uniform and may decrease or increase depending on the shape and properties of the clouds. In the environment, in addition to direct rays, also the reflected ones might be taken: the reflection is 80% on the snow (almost 100% if the snow is fresh and compact), 20% in the water, 17% on the sand and 3% on the grass. In addition, the water works as a lens and we must remember that if you are immersed up to 40–50 cm 5% of the rays affects us by reflection even on those parts of the skin (area under the chin, inside the arms, under the buttocks) which usually are not exposed. In addition, artificial UV exposure such as tanning beds and other devices that produce UVR should be avoided. The phototype indicates the ability to defend against the UVR that varies from individual to individual. It can be easily obtained taking into account the color of the complexion, the eyes and the hair and also the reaction of the skin to the sun exposure. The lower skin types (blond, red hair with fair skin that hardly tans) are those who do not adapt to sun exposure and are subjected to skin damage. It is also important to protect daily the skin and the HV recurrence zones with a sunscreen.

5.2 Sunscreens

Sunscreen is a lotion, spray, gel or other topical product that absorbs or reflects some of the sun's UVR and thus helps protect against sunlight.

Depending on the mode of action, sunscreens can be classified into physical sunscreens (i.e. those that reflect the sunlight) or chemical sunscreens (i.e. those that absorb the UVR). Chemical, organic sunscreens absorb over relatively narrow wavebands, mainly in the UVB but nowadays also extending into the UVA. Physical sunscreens are inorganic substances that reflect and scatter both UV and visible radiations [107]. Use of sunscreen can reduce chronic damaging and the carcinogenic effects of UV radiation and recurrence of cutaneous HV.

Currently, it is recommended to spread the sunscreen on the skin in two layers in such a way that its thickness is as close as possible to 2 mg/cm^2, which is what enables to achieve the expected SPF. In practice, however, only between 0.5 and 1.5 mg/cm^2 are used mostly because of the high price of sunscreen. The effect of application thickness is shown diagrammatically in 34 for the ideal scenario of uniform application. This demonstrates how light absorption depends strongly on thickness. For example, a sunscreen labeled SPF 16 is reduced to an SPF of 2 if the consumer applies 0.5 mg/cm^2. Uniformity of application is another related crucial factor. The same amount of sunscreen non-uniformly applied implies that some areas receive little or no sunscreen. In general, sun lotions should always be applied in abundant quantities, in 2 layers and repeated during the day, immediately after swimming and every 2 h if sweating occurs. Only after a few days it will it be possible to reduce the SPF of the cream used, once the skin has had the time to activate its defense systems. Use of sunscreen should never be interrupted, even once tanned, because the melanin filters 70% of the UVB but not the UVA and because over time its filtered capacity becomes less effective. The SPF is important in the choice of a solar product and the one suitable for each phototype should be increased if environmental conditions require it. The SPF is the ratio of the dose of UV radiation causing minimal erythema in unprotected skin to the dose which causes a minimal erythema in skin protected by the sunscreen. For example, if the normal MED is 30 mJ/cm^2 and the MED of the protected skin is 450 mJ/cm^2, the SPF is 15. In other words, application of sunscreen has caused an increase by a factor of 15 in the dose required to induce erythema. The SPF is principally a measure of the sunscreen UVB attenuation. Although conceptually very simple, it is often misunderstood by the general public, who think that using a high factor sunscreen will protect the skin against the harmful effects of UV radiation. Even if the sunscreen provides the protection indicated by the SPF, a day's exposure outdoors wearing a factor 15 sunscreen will still result in more than one MED for many individuals. This misconception often leads individuals to stay in the sun longer than they should. There is also a major difference between the highly controlled conditions in the sunscreen laboratory and outdoor real-life product use. [107]

6. Conclusion

Sunlight is the most common trigger in stimulating the HSV reactivation. It still not well known how the UVR determines the reactivation of the virus. Several hypotheses have been made but do not lead to a single common path with the other triggers. However, we know how to protect ourselves from solar radiation and what methods to use to avoid it or reduce its harmful effect on the skin.

Acknowledgements

The authors thank Angela Sabalic for her precious collaboration and the preparation of the figure and Giustina Casu for native language check and editing.

Conflict of interest

No conflict of interests is declared.

Author details

Vittorio Mazzarello*, Marco Ferrari, Stefano Decandia and Maria Alessandra Sotgiu
Skin Lab, Department of Biomedical Sciences, University of Sassari, Sassari, Italy

*Address all correspondence to: vmazza@uniss.it

References

[1] Pellet P, Roizman B. The family Herpesviridae: A brief introduction. In: Fields BN, Knipe DM, Howley PM, editors. Fields Virology. 5th ed. Lippincott Williams & Wilkins: Philadelphia; 2007. pp. 2479-2499

[2] Davison AJ, Eberle R, Ehlers B, Hayward GS, McGeoch DJ, Minson AC, et al. The order Herpesvirales. Archives of Virology. 2009;**154**(1):171-177

[3] Xu F, Schillinger JA, Sternberg MR, et al. Seroprevalence and coinfection with herpes simplex virus type 1 and type 2 in the United States, 1988-1994. The Journal of Infectious Diseases. 2002;**185**(8):1019-1024

[4] Gross G et al. How to manage recurrent orofacial herpes simplex virus-1 lesions. The Pharmaceutical Journal. 2009;**283**(7565):187-190

[5] Fatahzadeh M et al. Human herpes simplex virus infections: Epidemiology, pathogenesis, symptomatology, diagnosis, and management November. JAAD. 2007;**57**(5):737-763

[6] Farooq AV, Shukla D. Herpes simplex epithelial and stromal keratitis: An epidemiologic update. Survey of Ophthalmology. 2012;**57**(5):448-462

[7] Schmader K. Herpes Zoster. Annals of Internal Medicine. 2018;**169**(3):ITC19-ITC31

[8] Zak-Prelich M, Borkowski JL, Alexander F, Norval M. The role of solar ultraviolet irradiation in zoster. Epidemiology and Infection. 2002;**129**(3):593-597

[9] Brisson M, Edmunds WJ, Law B, Gay NJ, Walld R, Brownell M, et al. Epidemiology of varicella zoster infection in Canada and the United Kingdom. Epidemiology and Infection. 2001;**127**:305-314

[10] Agelidis AM, Shuckla D. Cell entry mechanisms of HSV: What we have learned in recent years. Future Virology. 2015;**10**(10):1145-1154

[11] Steiner I, Kennedy PG, Pachner AR. The neurotropic herpes viruses: Herpes simplex and varicella-zoster. Lancet Neurology. 2007;**6**:1015-1028

[12] Lee GE, Murray JW, Wolkoff AW, Wilson DW. Reconstitution of herpes simplex virus microtubule-dependent trafficking in vitro. Journal of Virology. 2006;**80**:4264-4275

[13] Saksena MM, Wakisaka H, Tijono B, Boadle RA, Rixon F, Takahashi H, et al. Herpes simplex virus type 1 accumulation, envelopment, and exit in growth cones and varicosities in mid-distal regions of axons. Journal of Virology. 2006;**80**:3592-3606

[14] Smith GA, Gross SP, Enquist LW. Herpesviruses use bidirectional fast-axonal transport to spread in sensory neurons. Proceedings of the National Academy of Sciences of the United States of America. 2001;**98**:3466-3470

[15] Oh M, Akhtar J, Desai P, Shukla D. A role for heparan sulfate in viral surfing. Biochemical and Biophysical Research Communications. 2010;**391**(1):176-181

[16] Shukla D, Spear PG. Herpesviruses and heparan sulfate: An intimate relationship in aid of viral entry. The Journal of Clinical Investigation. 2001;**108**(4):503-510

[17] Spear PG, Manoj S, Yoon M, Jogger CR, Zago A, Myscofski D. Different receptors binding to distinct interfaces on herpes simplex virus gD can trigger events leading to cell fusion and viral entry. Virology. 2006;**344**:17-24

[18] Clement C, Tiwari V, Scanlan PM, Valyi-Nagy T, Yue BYJT, Shukla D.

A novel role for phagocytosis-like uptake in HSV entry. The Journal of Cell Biology. 2006;**174**(7):1009-1021

[19] Suzik JB, Cliffe AR. Strength in diversity: Understanding the pathways of herpes simplex virus reactivation. Virology. 2018;**522**:81-91

[20] Knipe DM, Cliffe A. Chromatin control of herpes simplex virus lytic and latent infection. Nature Reviews Microbiology. 2008;**6**:211-221

[21] Deshmane SL, Fraser NW. During latency, herpes simplex virus type 1 DNA is associated with nucleosomes in a chromatin structure. Journal of Virology. 1989;**63**:943-947

[22] Kubat NJ, Tran RK, McAnany P, Bloom DC. Specific histone tail modification and not DNA methylation is a determinant of herpes simplex virus type 1 latent gene expression. Journal of Virology. 2004;**78**:1139-1149

[23] Wang Q-Y, Zhou C, Johnson KE, Colgrove RC, Coen DM, Knipe DM. Herpesviral latency-associated transcript gene promotes assembly of heterochromatin on viral lytic-gene promoters in latent infection. Proceedings of the National Academy of Sciences of the United States of America. 2005;**102**:16055-16059

[24] Kwiatkowski DL, Thompson HW, Bloom DC. The Polycomb group protein Bmi1 binds to the herpes simplex virus 1 latent genome and maintains repressive histone Marks during latency. Journal of Virology. 2009;**83**:8173-8181

[25] Nicoll MP, Hann W, Shivkumar M, Harman LER, Connor V, Coleman HM, et al. The HSV-1 latency-associated transcript functions to repress latent phase lytic gene expression and suppress virus reactivation from latently infected neurons. PLoS Pathogens. 2016;**12**:e1005539

[26] Cliffe AR, Garber DA, Knipe DM. Transcription of the herpes simplex virus latency-associated transcript promotes the formation of facultative heterochromatin on lytic promoters. Journal of Virology. 2009;**83**:8182-8190

[27] Raja P, Lee JS, Pan D, Pesola JM, Coen DM, Knipe DM. A Herpesviral lytic protein regulates the structure of latent viral chromatin. MBio. 2016;**7**:e00633-16-10

[28] Leib DA, Bogard CL, Kosz-Vnenchak M, Hicks KA, Coen DM, Knipe DM, et al. A deletion mutant of the latency-associated transcript of herpes simplex virus type 1 reactivates from the latent state with reduced frequency. Journal of Virology. 1989;**63**:2893-2900

[29] Trousdale MD, Steiner I, Spivack JG, Deshmane SL, Brown SM, MacLean AR, et al. In vivo and in vitro reactivation impairment of a herpes simplex virus type 1 latency-associated transcript variant in a rabbit eye model. Journal of Virology. 1991;**65**:6989-6993

[30] Nadelman CM, Newcomer VD. Herpes simplex virus infections. Postgraduate Medicine. 2000;**107**:189-200

[31] Woo SB, Challacombe SJ. Management of recurrent oral herpes simplex infections. Oral Surgery, Oral Medicine, Oral Pathology, Oral Radiology, and Endodontics. 2007;**103** (S12 Suppl):e1-e18

[32] Ichihashi M, Nagai H, Matsunaga K. Sunlight is an important causative factor of recurrent herpes simplex. Cutis. 2004;**74**(5 Suppl):14-18

[33] Ship II, Morris AL, Durocher RT, Burkett LW. Recurrent aphthous ulcerations and recurrent herpes labialis in a professional school student population. Oral Surgery, Oral Medicine, and Oral Pathology. 1960;**13**:1191-1202

[34] Cohen SR, Goodacre A, Lim S, Johnston J, Henssler C, Jeffers B, et al. Clinical outcomes and complications associated with fractional lasers: A review of 730 patients. Aesthetic Plastic Surgery. 2017;**41**(1):171-178

[35] Miller CS, Cunningham LL, Lindroth JE, Avdiushko SA. The efficacy of valacyclovir in preventing recurrent herpes simplex virus infections associated with dental procedures. Journal of the American Dental Association (1939). 2004;**135**:1311-1318

[36] Begolli Gerqari A, Ferizi M, Kotori M, Daka A, Hapciu S, Begolli I, et al. Activation of herpes simplex infection after tattoo. Acta Dermatovenerologica Croatica. 2018;**26**(1):75-76

[37] Aldea S, Joly L-M, Roujeau T, Oswald A-M, Devaux B. Postoperative herpes simplex virus encephalitis after neurosurgery: Case report and review of literature. Clinical Infectious Diseases. 2003;**36**:96-99

[38] Cohen F, Kemeny ME, Kearney KA, Zegans LS, Neuhaus JM, Conant MA. Persistent stress as a predictor of genital herpes recurrence. Archives of Internal Medicine. 1999;**159**:2430-2436

[39] Glaser R, Kiecolt-Glaser JK. Stress-associated immune modulation and its implications for reactivation of latent herpesviruses. In: Glaser R, Jones J, editors. Human Herpesvirus Infections. New York: Dekker; 1994. pp. 245-270

[40] Stowe RP, Peek MK, Cutchin MP, Goodwin JS. Herpesvirus reactivation and socioeconomic position: A community-based study. Journal of Epidemiology and Community Health. 2010;**64**:666-671

[41] Glaser R, Kiecolt-Glaser JK. Chronic stress modulates the virus-specific immune response to latent herpes simplex virus type 1. Annals of Behavioral Medicine. 1997;**19**(2):78-82

[42] Pazin G, Ho M, Jannetta P. Reactivation of herpes simplex virus after decompression of the trigeminal nerve root. The Journal of Infectious Diseases. 1978;**138**:405-409

[43] Wung PK, Holbrook JT, Hoffman GS, Tibbs AK, Specks U, Min YI, et al. Herpes zoster in immunocompromised patients: Incidence, timing, and risk factors. The American Journal of Medicine. 2005;**118**(12):1416-1418

[44] Shirtcliff EA, Coe CL, Pollak SD. Early childhood stress is associated with elevated antibody levels to herpes simplex virus type 1. Proceedings of the National Academy of Sciences of the United States of America. 2009;**106**(8):2963-2967

[45] Steptoe A, Shamaci-Tousi A, Gylfe A, Henderson B, Bergstrom S, Marmot M. Socioeconomic status, pathogen burden and cardiovascular risk. Heart. 2007;**93**:1567-1570

[46] Glaser R, Kiecolt-Glaser JK. Stress-induced immune dysfunction: Implications for health. Nature Reviews Immunology. 2005;**5**(3):243-251

[47] Gray J, Wreghitt T, Pavel P, Smyth R, Parameshwar J, Stewart S, et al. Epstein-Barr virus infection in heart and heart–lung transplant recipients: Incidence and clinical impact. The Journal of Heart and Lung Transplantation. 1995;**14**(4):640-646

[48] Marks JG, Miller JJ. Lookingbill and Marks' Principles of Dermatology. Philadelphia: Saunders Elsevier; 2013

[49] Baron ED, Suggs AK. Introduction to photobiology. Dermatologic Clinics. 2014;**32**(3):255-266

[50] Schaefer H, Moyal D, et al. Recent advances in sun protection. Seminars in Cutaneous Medicine and Surgery. 1998;**17**(4):266-275

[51] Bolognia JL, Schaffer JV, Duncan KO, et al, editors. Dermatology Essentials. Oxford: Saunders/Elsevier; 2014

[52] Lucas RM et al. Estimating the global disease burden due to ultraviolet radiation exposure. International Journal of Epidemiology. 2008;**6**(37):654-667

[53] Runger MT. Ultraviolet light. In: Bolognia JL, Jorizzo JL, Schaffer JV, editors. Dermatology. Philadelphia: Elsevier Saunders; 2012. pp. 1455-1465

[54] Schwarz T, Halliday GM. Photoimmunology. In: Lim HW, Honigsmann H, Hawk JLM, editors. Photodermatology. New York: Informa Healthcare USA; 2007. pp. 55-74

[55] Mehling A, Loser K, Varga G, et al. Overexpression of CD40 ligand in murine epidermis results in chronic skin inflammation and systemic autoimmunity. The Journal of Experimental Medicine. 2001;**194**:615-628

[56] Casciola-Rosen LA, Anhalt G, Rosen A. Autoantigens targeted in systemic lupus erythematosus are clustered in two populations of surface structures on apoptotic keratinocytes. The Journal of Experimental Medicine. 1994;**179**:1317-1330

[57] Elmets CA, Bergstresser PR, Tigelaar RE, Wood PJ, Streilein JW. Analysis of the mechanism of unresponsiveness produced by haptens painted on skin exposed to low dose ultraviolet radiation. The Journal of Experimental Medicine. 1982;**158**:781-794

[58] Schwarz T. 25 years of UV-induced immunosuppression mediated by T cells-from disregarded T suppressor cells to highly respected regulatory T cells. Photochemistry and Photobiology. 2008;**84**:10-18

[59] Schweintzger N, Gruber-Wackernagel A, Reginato E, Bambach I, Quehenberger F, Byrne SN, et al. Levels and function of regulatory T cells in patients with polymorphic light eruption: Relation to photohardening. The British Journal of Dermatology. 2015;**173**:519-526

[60] Schweintzger NA, Gruber-Wackernagel A, Shirsath N, Quehenberger F, Obermayer-Pietsch B, Wolf P. Influence of the season on vitamin D levels and regulatory T cells in patients with polymorphic light eruption. Photochemical & Photobiological Sciences. 2016;**15**:440-446

[61] Byrne SN, Beaugie C, O'sullivan C, Leighton S, Halliday GM. The immune-modulating cytokine and endogenous Alarmin interleukin-33 is upregulated in skin exposed to inflammatory UVB radiation. The American Journal of Pathology. 2011;**179**:211-222

[62] Toews GB, Bergstresser PR, Streilein JW. Epidermal Langerhans cell density determines whether contact hyper sensitivity or unresponsiveness follows skin painting with DNFB. Journal of Immunology. 1980;**124**:445-453

[63] Noonan FP, Bucana C, Sauder DN, DeFabo EC. Mechanism of systemic immunesuppression by UV irradiation in vivo. II. The UV effects on number and morphology of epidermal Langerhans cells and the UV-induced suppression of contact hypersensitivity have different wavelength dependencies. Journal of Immunology. 1984;**132**:2408-2416

[64] Hart PH, Grimbaldeston MA, Finlay-Jones JJ. Sunlight, immunosuppression and skin cancer: Role of histamine and mast cells. Clinical and Experimental Pharmacology & Physiology. 2001;**28**:1-8

[65] van der Molen RG, Out-Luiting C, Claas FH, et al. Ultraviolet-B radiation induces modulation of antigen presentation of herpes simplex virus by human epidermal cells. Human Immunology. 2001;**62**(6):589-597

[66] Norval M. The effect of ultraviolet radiation on human viral infections. Photochemistry and Photobiology. 2006;**82**(6):1495-1504

[67] Patra V, Byrne SN, Wolf P. The skin microbiome: Is it affected by UV-induced immune suppression? Front Microbiol. 2016;**10**(7):1235

[68] Termorshuizen F, Garssen J, Norval M, Koulu L, Laihia J, Leino L, et al. A review of studies on the effects of ultraviolet irradiation on the resistance to infections: Evidence from rodent infection models and verification by experimental and observational human studies. International Immunopharmacology. 2002;**2**(2-3):263-275

[69] Spruance SL. Herpes simplex labialis. In: Sacks SL, Straus SE, Whitley RJ, Griffiths PD, editors. Clinical Management of Herpes Viruses. 4th ed. Amsterdam: IOS Press; 1995. pp. 11-20

[70] Rooney JF, Bryson Y, Mannix ML, et al. Prevention of ultraviolet-light-induced herpes labialis by sunscreen. Lancet. 1991;**338**(8780):1419-1422

[71] Duthie MS, Kimber I, Norval M. The effects of ultraviolet radiation on the human immune system. The British Journal of Dermatology. 1999;**140**:995-1009

[72] Laihia JK, Jansen CT. Solar-simulating ultraviolet irradiation of the skin of human subjects in vivo produces Langerhans cell responses distinct from irradiation ex vivo and in vitro. Experimental Dermatology. 2000;**9**:240-247

[73] Termorshuizen F, Garssen J, Norval M, et al. A review of studies on the effects of ultraviolet irradiation on the resistance to infections: Evidence from rodent infection models and verification by experimental and observational humane studies. International Immunopharmacology. 2002;**2**:263-275

[74] Perna JJ, Mannix ML, Rooney JF, Notkins AL, Straus SE. Reactivation of latent herpes simplex virus infection by ultraviolet light: A human model. Journal of the American Academy of Dermatology. 1987;**17**:473-478

[75] Norval M, El-Ghorr AA. UV radiation and mouse models of herpes simplex virus infection. Photochemistry and Photobiology. 1996;**64**(2):242-245

[76] Laycock KA, Lee SF, Brady RH, et al. Characterization of a murine model of recurrent herpes simplex viral keratitis induced by ultraviolet B radiation. Investigative Ophthalmology & Visual Science. 1991;**32**(10):2741-2746

[77] James W et al. Andrews' Diseases of the Skin Clinical Dermatology. 11th ed. Philadelphia: Saunders-Elsevier; 2011

[78] Spruance SL, Freeman DJ, Stewart JC, McKeough MB, Wenerstrom LG, Krueger GG, et al. The natural history of ultraviolet radiation-induced herpes simplex labialis and response to therapy with peroral and topical formulations of acyclovir. The Journal of Infectious Diseases. 1991;**163**:728-734

[79] Duteil L, Queille-Roussel C, Loesche C, Verschoore M. Assessment of the effect of a sunblock stick in the prevention of solar-simulating ultraviolet light-induced herpes labialis. Journal of Dermatological Treatment. 1998;**9**(1):11-14

[80] Ludema C, Cole SR, Poole C, Smith JS, Schoenbach VJ, Wilhelmus KR. Association between unprotected

ultraviolet radiation exposure and recurrence of ocular herpes simplex virus. American Journal of Epidemiology. 2014;**179**(2):208-215

[81] Gallerani M, Manfredini R. Seasonal variation in herpes zoster infection. The British Journal of Dermatology. 2000;**142**:588-589

[82] Young TB et al. Cross-sectional study of recurrent herpes labialis. Prevalence and risk factors. American Journal of Epidemiology. 1988;**127**:612-625

[83] Axell T et al. Occurrence of recurrent herpes labialis in an adult Swedish population. Acta Odontologica Scandinavica. 1990;**48**:119-123

[84] Reichart PA. Oral mucosal lesions in a representative cross-sectional study of aging Germans. Community Dentistry and Oral Epidemiology. 2000;**28**:390-398

[85] Norval M, el-Ghorr A, Garssen J, Van Loveren H. The effects of ultraviolet light irradiation on viral infections. The British Journal of Dermatology. 1994;**130**(6):693-700

[86] Openshaw H, Asher LVS, Wohlenberg C, Sekizawa T, Notkins AL. Acute and latent infection of sensory ganglia with herpes simplex virus. Immune control and virus reactivation. The Journal of General Virology. 1979;**44**:205-215

[87] Lycke E, Kristensson K, Svennerholm B, Vahine A, Ziegler R. Uptake and transport of herpes simplex virus in neurites of rat dorsal root ganglia cells in culture. The Journal of General Virology. 1984;**65**:55-64

[88] Blank H, Haines H. Experimental human reinfection with herpes simplex virus. The Journal of Investigative Dermatology. 1973;**61**:223-225

[89] Chida Y, Mao X. Does psychosocial stress predict symptomatic herpes simplex virus recurrence? A meta-analytic investigation on prospective studies. Brain, Behavior, and Immunity. 2009;**23**:917-925

[90] El Hayderi L, Delvenne P, Rompen E, Senterre JM, Nikkels AF. Herpes simplex virus reactivation and dental procedures. Clinical Oral Investigations. 2013;**17**:1961-1964

[91] Padgett DA, Sheridan JF, Dorne J, Berntson GG, Candelora J, Glaser R. Social stress and the reactivation of latent herpes simplex virus type 1. Proceedings of the National Academy of Sciences of the United States of America. 1998;**95**:7231-7235

[92] Loiacono CM, Taus NS, Mitchell WJ. The herpes simplex virus type 1 ICP0 promoter is activated by viral reactivation stimuli in trigeminal ganglia neurons of transgenic mice. Journal of Neurovirology. 2003;**9**:336-345

[93] Henderson G, Peng W, Jin L, et al. Regulation of caspase 8- and caspase 9-induced apoptosis by the herpes simplex virus type 1 latency-associated transcript. Journal of Neurovirology. 2002;**8**(suppl 2):103-111

[94] Stefanato CM, Yaar M, Bhawan J, Phillips TJ, Kosmadaki MG, Botchkarev V, et al. Modulations of nerve growth factor and Bcl-2 in ultraviolet-irradiated human epidermis. Journal of Cutaneous Pathology. 2003;**30**:351-357

[95] Wilcox CL, Johnson EM. Nerve growth factor deprivation results in the reactivation of latent herpes simplex virus in vitro. Journal of Virology. 1987;**61**:2311-2315

[96] Hill JM, Garza HH, Helmy MF, Cook SD, Osborne PA, Johnson EM, et al. Nerve growth factor antibody

stimulates reactivation of ocular herpes simplex virus type 1 in latently infected rabbits. Journal of Neurovirology. 1997;**3**:206-211

[97] Camarena V, Kobayashi M, Kim JY, Roehm P, Perez R, Gardner J, et al. Nature and duration of growth factor signaling through receptor tyrosine kinases regulates HSV-1 latency in neurons. Cell Host & Microbe. 2010;**8**:320-330

[98] Cliffe AR, Arbuckle JH, Vogel JL, Geden MJ, Rothbart SB, Cusack CL, et al. Neuronal stress pathway mediating a histone methyl/phospho switch is required for herpes simplex virus reactivation. Cell Host & Microbe. 2015;**18**:649-658

[99] Kobayashi M, Wilson AC, Chao MV, Mohr I. Control of viral latency in neurons by axonal mTOR signaling and the 4E-BP translation repressor. Genes & Development. 2012;**26**:1527-1532

[100] Linderman JA, Kobayashi M, Rayannavar V, Fak JJ, Darnell RB, Chao MV, et al. Immune escape via a transient gene expression program enables productive replication of a latent pathogen. Cell Reports. 2017;**18**:1312-1323

[101] Du T, Zhou G, Roizman B. Induction of apoptosis accelerates reactivation of latent HSV-1 in ganglionic organ cultures and replication in cell cultures. Proceedings of the National Academy of Sciences of the United States of America. 2012;**109**:14616-14621

[102] Messer HGP, Jacobs D, Dhummakupt A, Bloom DC. Inhibition of H3K27me3-specific histone demethylases JMJD3 and UTX blocks reactivation of herpes simplex virus 1 in trigeminal ganglion neurons. Journal of Virology. 2015;**89**:3417-3420

[103] Workman A, Eudy J, Smith L, da Silva LF, Sinani D, Bricker H, et al. Cellular transcription factors induced in trigeminal ganglia during dexamethasone-induced reactivation from latency stimulate bovine herpesvirus 1 productive infection and certain viral promoters. Journal of Virology. 2012;**86**:2459-2473

[104] Mills J, Hauer L, Gottlieb A, Dromgoole S, Spruance S. Recurrent herpes labialis in skiers: Clinical observations and effect of sunscreen. The American Journal of Sports Medicine. 1987;**15**:76-78

[105] Worrall G. Herpes labialis. BMJ Clinical Evidence. 2009;**2009**:1704

[106] Mazzarello V, Ferrari M, Piu G, Pomponi V, Solinas G. Do sunscreen prevent recurrent herpes labialis in summer? Journal of Dermatological Treatment. 2018;**23**:1-4

[107] Moseley H. Photoprotection. In: Ferguson J, Dover JS, editors. Photodermatology. 1st ed. London: Manson publishing; 2006. pp. 21-28

Section 3

Varicella-Zoster Virus

Chapter 3

Neurologic Complications of Varicella-Zoster Virus Infection

Hideto Nakajima, Makoto Hara, Akihiko Morita and Satoshi Kamei

Abstract

Varicella-zoster virus (VZV) causes a diverse spectrum of neurologic complications: aseptic meningitis, encephalitis, cerebral infarction associated with granulomatous vasculitis, myelitis, and cranial polyneuropathy. These VZV-associated central nervous system (CNS) diseases usually result from reactivation of latent infection in immunosuppressive conditions, such as old age, diabetes mellitus, cancer, human immunodeficiency virus (HIV) infection, and the use of immunosuppressive drugs. However, they also occur in immunocompetent subjects. Since VZV antigen or DNA is often detected in the cerebrospinal fluid of these patients, it is thought that reactivated VZV reaches the central nervous system by direct spread from latently infected sensory ganglia. Analysis of cerebrospinal fluid by PCR is important for the diagnosis of VZV-associated CNS diseases particularly in the absence of exanthema/herpes zoster. Clinicians should be aware of the neurologic complications of VZV infection, because early acyclovir therapy is necessary for these disorders.

Keywords: varicella-zoster virus (VZV), neurologic infections, central nervous system (CNS), encephalitis, meningitis

1. Introduction

The clinical manifestations of varicella-zoster virus (VZV) infections of the central nervous system (CNS) include aseptic meningitis, encephalitis, cerebral infarction associated with granulomatous vasculitis, myelitis, and multiple cranial neuropathies (**Figure 1**) [1–4]. In these patients, viral antigens or DNA are often detected in the cerebrospinal fluid (CSF) or the sites of pathology. Thus, those neurological disorders reflect reactivation of latent VZV in the trigeminal ganglia and dorsal root ganglion, with subsequent spread of the infection into the CNS [1]. In addition, the incidence of CNS complications caused by VZV is more likely higher in elderly individuals; those with underlying diseases, such as malignant tumors and HIV; and those who are immunosuppressed due to the use of steroids or immunosuppressive drugs [5–8]. However, it can also affect healthy individuals; therefore, these CNS VZV infections may be suspected even in patients without underlying diseases. Among CNS infections caused by VZV, diseases other than meningitis are rare; however, clinicians should be aware of the various clinical features of CNS infections caused by VZV to start early and accurate antivirus drugs for treatment.

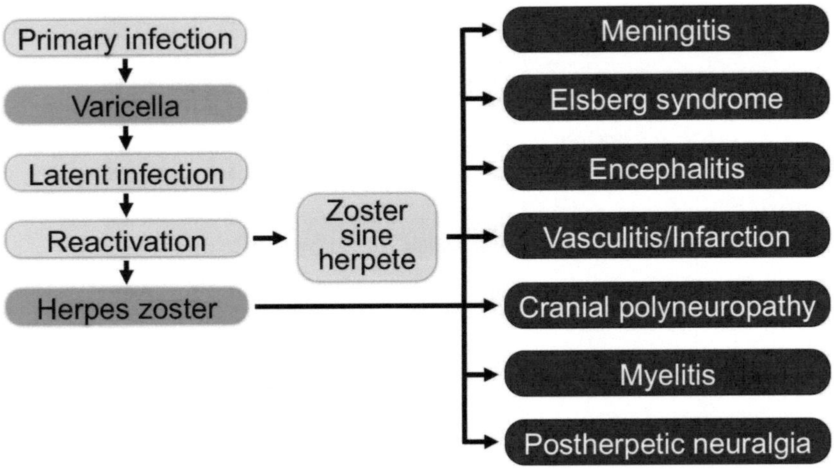

Figure 1.
Neurological complications associated with VZV reactivation.

2. Zoster sine herpete

If neurological symptoms occur simultaneously or around the same time as the onset of herpes zoster infection, they can be considered complications of VZV infections. However, if lesions such as rashes, shingles, or blisters are not observed, VZV as a causative agent is likely to be missed. In fact, because VZV can cause CNS complications in the absence of skin lesions, such cases are referred to as zoster sine herpete [9]; clinicians have to consider the possibility of VZV as a causative virus in patients with neurological infections, such as meningitis or encephalitis (**Figure 1**).

3. Aseptic meningitis

Meningitis is inflammation of the pia mater and the arachnoid that cover the surface of the brain. Its clinical signs include fever, headache, nausea, vomiting, and meningeal irritation symptoms, such as nuchal rigidity and Kernig's sign. Furthermore, jolt accentuation and neck flexion tests are often positive. However, these are common symptoms and findings of meningitis regardless of the cause. The CSF examination shows monocyte-dominant pleocytosis and elevated protein levels with normal glucose levels. Patients with meningitis wherein bacteria are not detected via the CSF test are generally diagnosed as having aseptic meningitis. Most cases of aseptic meningitis involve viral meningitis. The most common virus that causes viral meningitis is enterovirus. In adults, enterovirus is followed by herpes simplex virus type 2 (HSV-2) and VZV [10], and VZV infection accounts for 8% of the total meningitis cases [11].

VZV meningitis can sometimes cause cranial polyneuropathy or dysuria due to sacral radiculopathy, and dysuria due to sacral radiculopathy is known as Elsberg syndrome. Meningitis caused by VZV is also frequently observed among healthy young individuals. Such a condition generally has a good prognosis and rarely causes any sequelae.

4. Elsberg syndrome

Elsberg syndrome is caused by bilateral sacral radiculopathy, which is characterized by urinary retention, sensory disturbance, and neuralgia of the perineum and lower limbs. Although Elsberg syndrome was originally characterized by urinary retention due to sacral radiculopathy associated with genital herpes, it is now defined as aseptic meningitis-associated sacral radiculopathy. As the causative virus, HSV, particularly HSV-2, is the most common cause, followed by VZV [12, 13]. When urinary retention occurs, urethral catheterization is required. However, this condition resolves as meningitis improves.

4.1 Case 1: Elsberg syndrome

A 32-year-old man was admitted to our hospital because of high fever, headache, nausea, acute urinary retention, and dysesthesia in a lumbosacral dermatome distribution. There were no motor symptoms and no rash. CSF analysis showed 249 leukocytes/mm^3, 70 mg/dl protein, and positive of VZV DNA by PCR. Gadolinium-enhanced MRI revealed the meningeal lesions of the conus medullaris and the swollen radicular fibers in the upper lumbar spinal canal. Treatment of acyclovir and dexamethasone for 2 weeks led to complete resolution of meningitis and urinary retention.

5. Encephalitis and cerebral infarction associated with granulomatous vasculitis

The symptoms of encephalitis include acute disturbance of consciousness, headache, fever, and convulsions. Neurological findings of encephalitis include meningeal irritation symptoms, such as nuchal rigidity; however, patients with encephalitis sometimes present with motor paralysis and sensory disturbance due to parenchymal brain damage.

Among the pathogens that cause viral encephalitis, VZV is the second most common cause following HSV, accounting for 5% of the total encephalitis cases [14]. According to a recent analysis that used PCR, though, the risk of VZV encephalitis increases in elderly individuals, those with herpes zoster ophthalmicus, and those with disseminated herpes zoster, and this result indicates that the incidence of VZV encephalitis might have increased [15].

The clinical manifestations of VZV encephalitis include meningoencephalitis and vasculopathy [16]. The meningoencephalitis form shows no detectable lesions on MRI. In contrast, the vasculopathy form is characterized by non-specific ischemia, hemorrhagic lesions, and multiple white matter lesions on MRI [16]. Pathological studies suggested that VZV encephalitis develops based on vasculopathy in the large and small vessels. Therefore, MRI typically demonstrates ischemic or hemorrhagic infarction in both gray and white matter and particularly at gray-white matter junctions as characteristic imaging findings of VZV encephalitis [16]. In VZV encephalitis, lesions in the temporal lobe and limbic system, which are often observed in patients with herpes simplex encephalitis, are rare. Moreover, hemorrhagic lesions and necrosis, which are characteristics of herpes simplex encephalitis, are not commonly observed. Because VZV DNA is generally detected in the CSF of adult patients with VZV encephalitis, direct viral invasion to the CNS is believed to be the pathology of VZV encephalitis. In contrast, in varicella encephalitis in children who develop acute cerebellar ataxia associated with varicella infection, VZV is not detected in the CSF. Therefore, a secondary immunological allergic mechanism is considered as the pathology of varicella encephalitis.

Cerebral infarction caused by granulomatous vasculitis is a complication of herpes zoster infection [1, 17]. A typical patient presents with herpes zoster ophthalmicus, followed by postherpetic contralateral hemiplegia, and develops cerebral infarction between the eighth day and sixth month after herpes zoster infection (average of 7 weeks) [18, 19]. Patients with cerebral infarction often present with stenosis or obstruction in the anterior cerebral artery or middle cerebral artery. Because VZV DNA and antigens are detected in the walls of cerebral arteries, this evidence should provide an anatomic pathway for transaxonal spread of VZV after reactivation from trigeminal ganglia as a mechanism of intracerebral VZV vasculopathy [20–22]. The incidence of stroke increases 6 months after the onset of herpes zoster infection [23], and VZV vaccine and antiviral drug therapy may help reduce the risk of stroke after herpes zoster infection [24]. Cerebral infarction can also develop after varicella infection in children [25]. Although it is rare, it occurs within 6 months after varicella infection, and a similar mechanism as cerebral infarction after varicella zoster infection is considered [25]. In these conditions, VZV, which causes latent infection in the trigeminal ganglion after varicella infection, reactivates and directly invades the vessels in the CNS.

5.1 Case 2: meningoencephalitis

The patient was a 77-year-old woman who was admitted to our hospital due to convulsions and impaired consciousness. She presented with a Glasgow Coma Scale score of E1V1M4, and positive nuchal rigidity was observed. The convulsions were treated with the intravenous injection (IV) of diazepam and intramuscular injection of phenobarbital. However, the patient had high fever after admission at the hospital. CSF examination showed increased cell count (125.0 mg/dl), elevated protein level (125.0 mg/dl), and positivity for VZV DNA, and she was then diagnosed with VZV infection. The patient was treated with acyclovir and dexamethasone, and she regained consciousness and was able to talk on the second day of hospitalization. On the seventh day, she recovered with lucid consciousness without sequelae (**Figure 2**). Her MRI showed no abnormal lesions in the brain parenchyma, and she was diagnosed with meningoencephalitis.

5.2 Case 3: cerebral infarction associated with granulomatous vasculitis

The patient was a 76-year-old man who developed infarction in the right medial hypothalamus 34 days after the onset of right ophthalmic herpes zoster. He further

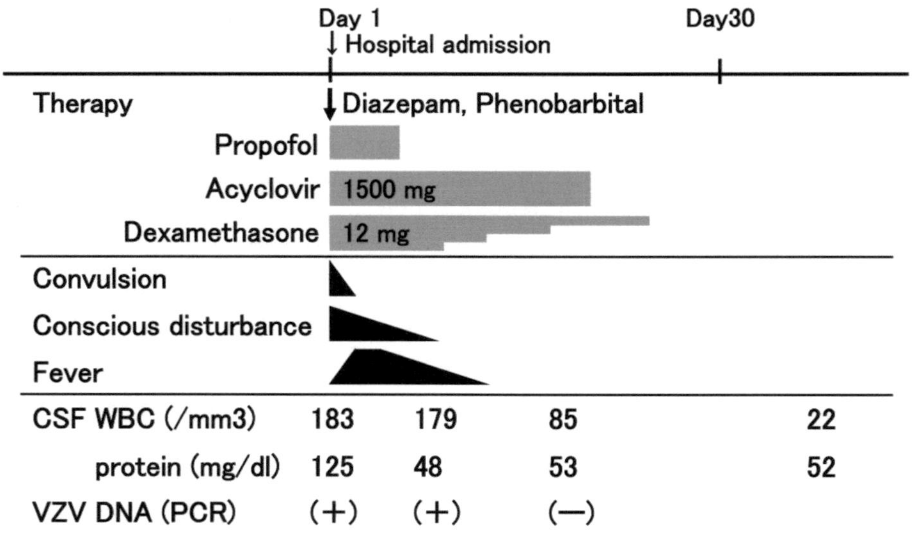

Figure 2.
Clinical course (Case 1).

developed an infarction in the right occipital lobe 73 days after the onset of herpes zoster infection. Although the MRI obtained while the patient presented with herpes zoster rash did not show any abnormal findings, the MRI performed 73 days later showed severe stenosis of the posterior communicating artery.

5.3 Case 4: cerebral infarction associated with granulomatous vasculitis

The patient was a 52-year-old woman with systemic lupus erythematosus (SLE) who exhibited altered levels of consciousness during immunotherapy for SLE. The CSF test showed pleocytosis, an elevated protein level, and positivity for VZV DNA, and the patient was then diagnosed with VZV meningoencephalitis. Brain MRI showed cerebral infarction in the left cerebral white matter, and MR angiogram showed stenosis of the left middle cerebral artery.

6. Cranial polyneuropathy

Cranial nerve palsy can sometimes develop in patients with herpes zoster of face or neck regions. Facial nerve palsy accompanying herpes zoster infection is known as Ramsay Hunt syndrome, and those patients often exhibit cranial polyneuropathy [26, 27]. Lower cranial polyneuropathy causes dysphagia, dysarthria, and hoarseness. Furthermore, there was no elevation or constriction in the unilateral soft palate, and tongue deviation and muscular weakness of the sternocleidomastoid and trapezius muscles were observed due to unilateral glossopharyngeal, vagus, accessary, and hypoglossal nerve paralyzes. Cranial polyneuropathy is often accompanied by meningitis, and CSF examination showed pleocytosis and elevated protein levels. In most cases, brain MRI shows no abnormalities. However, contrast MRI sometimes shows enhancement in the affected cranial nerves. As a mechanism of this condition, reactivation of VZV from the geniculate ganglion could result inflammatory process, circulatory disturbance, or edema to involve cranial nerves [28].

6.1 Case 5: lower cranial polyneuropathy

A 64-year-old woman developed acute paralysis of the IX, X, XI, and XII nerves on the left side after experiencing pain in the left ear and throat. CSF examination revealed lymphocytic pleocytosis and elevated protein levels. VZV DNA was detected with PCR using CSF. She was diagnosed with cranial polyneuropathy due to VZV reactivation. After the oral administration of antiviral agent and steroid, all signs and symptoms dramatically improved. Notably, there was no evidence of cutaneous or mucosal rash during the entire course of the disease. VZV reactivation should be included in the differential diagnosis of multiple cranial nerve palsies, particularly with pain and even without rash.

6.2 Case 6: lower cranial polyneuropathy

The patient was a 66-year-old man who presented with dysphagia and hoarseness 2 days after the onset of pain in the left occipital region to the shoulder. At an otorhinolaryngology clinic, recurrent nerve paralysis was observed, and lesions of herpes zoster were noted in the left side of the neck. Left glossopharyngeal, vagal, accessory, and hypoglossal nerve paralyzes were observed during neurological examination. CSF examination showed increased cell count and positivity for VZV DNA, and the patient was diagnosed with multiple lower cranial polyneuropathy.

7. Myelitis

VZV myelitis is a rare clinical manifestation. However, Brown-Séquard syndrome and transverse myelopathy may occur as a complication of herpes zoster infection [6, 7, 29–31]. Previous reports revealed that myelitis occurred in elderly or immunocompromised patients, such as those with HIV infection, and this condition often had severe sequelae such as motor paralysis [6, 7]. MRI shows low to equal signal intensity on T1-weighted images and high signal intensity on T2-weighted images, and sometimes, contrast enhancement can be observed as spinal cord lesions. In addition, MRI enhancement may be observed not only in the lesions in the spinal cord but also in the meninges around the spinal cord as well as in the dorsal root nerve, and these findings indicate myeloradiculitis.

Myelitis is often characterized by myelopathy at a level consistent with the spinal segment affected by herpes zoster. As a pathogenesis of this condition, reactivated VZV in the dorsal root ganglion of the spinal cord directly invades the spinal cord from the dorsal root nerve, resulting in myelitis [6, 7]. Moreover, spinal cord lesions are caused by vasculopathy, such as damage to the anterior spinal artery due to vasculitis, similar to cerebral infarction after herpes zoster infection, and this may be considered another mechanism.

7.1 Case 7: myelitis

The patient was a 60-year-old man with right lower extremity paralysis and sensory disturbance of the right trunk and lower extremity who was diagnosed with VZV myelitis based on CSF examination. MRI of the spinal cord showed a high-signal lesion in the right posterior funiculus at the thoracic vertebral level of Th6 and Th7, which should indicate that VZV directly invaded the spinal cord from the dorsal root.

7.2 Case 8: myelitis

An 87-year-old woman developed weakness of the right lower limb 2 days after developing herpes zoster lesions in the right side of the chest. Neurological examination revealed a spastic palsy in the right lower limb and loss of pain and temperature sensation in the left side to T6. However, vibration and position senses were not impaired in both sides. Thus, the patient presented with incomplete Brown-Séquard syndrome. Spinal T2-weighted MRI images showed a high-intensity lesion in the right side of the spinal cord except at the posterior funiculus at the Th2 level. CSF analysis showed the following results: leukocyte count, 109/mm^3, and protein level, 79 mg/dl, as well as negativity for VZV PCR, elevated titer levels for anti-VZV IgM and IgG, and increased IgG index. Although she was treated with a combination of acyclovir and steroid pulse therapy, her weakness in the right lower limb did not improve. In this case, because the posterior funiculus circulating from the posterior spinal artery was not involved, the incomplete Brown-Séquard syndrome may have been caused by spinal cord infarction due to VZV vasculitis of the anterior spinal artery.

8. Postherpetic neuralgia

Although most cases of acute herpes zoster are self-limited, about 10–15% of patients with herpes zoster will develop postherpetic neuralgia (PHN) [32], particularly in older adults [33]. Immunosuppressed patients have a higher incidence of PHN. PHN refers to pain persisting for months to years after the resolution of the rash. Sensory symptoms can include pain, numbness, dysesthesias, and allodynia (pain precipitated by movement) in the affected dermatome. And these symptoms may be severe enough to restrict sleep, appetite, or daily activities. The diagnosis of PHN is clear-cut and could be made if those sensory symptoms including pain persist beyond 4 months in the same distribution as a preceding episode of acute herpes zoster [34]. Gabapentin, pregabalin, tricyclic antidepressants, and opioids are generally the first-line drugs for the treatment of PHN [35–37]. Vaccines are also available for prevention of acute zoster and PHN [38, 39].

9. Diagnosis

For the diagnosis of CNS infection caused by VZV, the detection of VZV DNA with PCR using CSF is necessary [40–42]. However, a negative VZV DNA result does not rule out VZV infection, and particularly, PCR examination after the initiation of antiviral treatment will likely turn out negative. Thus, testing should be conducted using CSF before the antiviral treatment. When measuring anti-VZV antibodies, a significant increase of the anti-VZV antibody titer in CSF over the course of the illness or findings suggesting the production of intrathecal antibody [serum/CSF antibody ratio $\leq$ 20 or antibody titer index = (CSF antibody/serum antibody)/(CSF albumin/serum albumin $\geq$ 2)] should be confirmed.

10. Therapy

Antiviral therapy with intravenous acyclovir (10 mg/kg intravenous every 8 hours) should be initiated as soon as the diagnosis is considered [43]. Although the prognosis of meningitis is good, encephalitis and myelitis often result in sequelae, and a delay in the initiation of acyclovir treatment leads to poor prognosis. Therefore, if CNS infection caused by VZV is suspected based on clinical symptoms as well as CSF examination and imaging findings, the administration of acyclovir must be immediately initiated. According to a recent study from the UK, some causative pathogens were identified in 42% of acute encephalitis cases, of which HSV or VZV was identified in one-fourth of the cases. Thus, the administration of acyclovir should be immediately initiated if encephalitis is clinically suspected, and then, acyclovir should be administered for 2 weeks in an immunocompetent host and for 3 weeks in an immunosuppressive host if encephalitis caused by HSV or VZV is confirmed [5]. In CNS infection caused by VZV, the standard administration period is similar. There is no evidence showing the therapeutic effect of adjunctive corticosteroid use. However, corticosteroids suppress the inflammatory response accompanied by cytotoxicity due to the host immune response to viral infection, and in cases of encephalitis/vasculitis, myelitis, and cranial polyneuropathy, the adjunctive administration of dexamethasone or steroid pulse therapy with acyclovir is recommended.

11. Conclusions

VZV causes the diverse spectrum of neurologic complications: aseptic meningitis, encephalitis, cerebral infarction associated with granulomatous vasculitis, myelitis, and cranial polyneuropathy. Clinicians should be aware of the neurologic complications of VZV, because early acyclovir therapy is necessary for these disorders.

Acknowledgements

The authors would like to express their gratitude to Dr. Mitsuru Matsuki (Radiology, Kindai University), Dr. Takafumi Hosokawa (Neurology, Osaka Medical College), Dr. Shinichi Haginomori (Otorhinolaryngology, Osaka Medical College), Dr. Hiroshi Ymamamoto (Radiology, Sumitomo Hospital), and Dr. Mitsuyosi Ayabe and Dr. Kazuhito Noda (Neurology, Kurume University) for providing the information of the cases.

This work was supported by JSPS KAKENHI, Grant Number 16 K09704, from the Ministry of Education, Culture, Sports, Science and Technology, Japan.

Conflict of interest

The authors declare no conflicts of interest.

Author details

Hideto Nakajima*, Makoto Hara, Akihiko Morita and Satoshi Kamei
Division of Neurology, Department of Internal Medicine, Nihon University School of Medicine, Tokyo, Japan

*Address all correspondence to: nakajima.hideto@nihon-u.ac.jp

References

[1] Echevarría JM, Casas I, Martínez-Martín P. Infections of the nervous system caused by varicella-zoster virus: A review. Intervirology. 1997;**40**:72-84. DOI: 10.1159/000150535

[2] Straus SE, Ostrove JM, Inchauspé G, Felser JM, Freifeld A, Croen KD, et al. NIH conference. Varicella-zoster virus infections. Biology, natural history, treatment, and prevention. Annals of Internal Medicine. 1988;**108**(2):221-237

[3] Oxman MN. Immunization to reduce the frequency and severity of herpes zoster and its complications. Neurology. 1995;**45**(12 Suppl 8):S41-S46

[4] Galil K, Choo PW, Donahue JG, Platt R. The sequelae of herpes zoster. Archives of Internal Medicine. 1997;**157**(11):1209-1213

[5] Gnann JW Jr. Varicella-zoster virus: Atypical presentations and unusual complications. The Journal of Infectious Diseases. 2002;**186**(Suppl 1):S91-S98. DOI: 10.1086/342963

[6] Manian FA, Kindred M, Fulling KH. Chronic varicella-zoster virus myelitis without cutaneous eruption in a patient with AIDS: Report of a fatal case. Clinical Infectious Diseases. 1995;**21**(4):986-988

[7] Chrétien F, Gray F, Lescs MC, Geny C, Dubreuil-Lemaire ML, Ricolfi F, et al. Acute varicella-zoster virus ventriculitis and meningo-myelo-radiculitis in acquired immunodeficiency syndrome. Acta Neuropathologica. 1993;**86**(6):659-665

[8] Glesby MJ, Moore RD, Chaisson RE. Clinical spectrum of herpes zoster in adults infected with human immunodeficiency virus. Clinical Infectious Diseases. 1995;**21**(2):370-375

[9] Gilden DH, Kleinschmidt-DeMasters BK, LaGuardia JJ, Mahalingam R, Cohrs RJ. Neurologic complications of the reactivation of varicella-zoster virus. The New England Journal of Medicine. 2000;**342**(9):635-645. DOI: 10.1056/NEJM200003023420906

[10] Ihekwaba UK, Kudesia G, McKendrick MW. Clinical features of viral meningitis in adults: Significant differences in cerebrospinal fluid findings among herpes simplex virus, varicella zoster virus, and enterovirus infections. Clinical Infectious Diseases. 2008;**47**(6):783-789. DOI: 10.1086/591129

[11] Kupila L, Vuorinen T, Vainionpää R, Hukkanen V, Marttila RJ, Kotilainen P. Etiology of aseptic meningitis and encephalitis in an adult population. Neurology. 2006;**66**:75-80. DOI: 10.1212/01.wnl.0000191407.81333.00

[12] Eberhardt O, Küker W, Dichgans J, Weller M. HSV-2 sacral radiculitis (Elsberg syndrome). Neurology. 2004;**63**(4):758-759

[13] Abe M, Araoka H, Kimura M, Yoneyama A. Varicella zoster virus meningoencephalitis presenting with Elsberg syndrome without a rash in an immunocompetent patient. Internal Medicine. 2015;**54**(16):2065-2067. DOI: 10.2169/internalmedicine.54.4341

[14] Granerod J, Ambrose HE, Davies NW, Clewley JP, Walsh AL, Morgan D, et al. Causes of encephalitis and differences in their clinical presentations in England: A multicentre, population-based prospective study. The Lancet Infectious Diseases. 2010;**10**:835-844. DOI: 10.1016/S1473-3099(10)70222-X

[15] Braun-Falco M, Hoffmann M. Herpes zoster with progression to acute varicella zoster virus-meningoencephalitis. International

Journal of Dermatology. 2009;**48**(8):834-839

[16] Gilden D, Cohrs RJ, Mahalingam R, Nagel MA. Varicella zoster virus vasculopathies: Diverse clinical manifestations, laboratory features, pathogenesis, and treatment. Lancet Neurology. 2009;**8**:731-740. DOI: 10.1016/S1474-4422(09)70134-6

[17] McKelvie PA, Collins S, Thyagarajan D, Trost N, Sheorey H, Byrne E. Meningoencephalomyelitis with vasculitis due to varicella zoster virus: A case report and review of the literature. Pathology. 2002;**34**:88-93

[18] Reshef E, Greenberg SB, Jankovic J. Herpes zoster ophthalmicus followed by contralateral hemiparesis: Report of two cases and review of literature. Journal of Neurology, Neurosurgery, and Psychiatry. 1985;**48**:122-127

[19] Ahmad NM, Boruchoff SE. Multiple cerebral infarcts due to varicella-zoster virus large-vessel vasculopathy in an immunocompetent adult without skin involvement. Clinical Infectious Diseases. 2003;**37**:e16-e18

[20] Melanson M, Chalk C, Georgevich L, Fett K, Lapierre Y, Duong H, et al. Varicella-zoster virus DNA in CSF and arteries in delayed contralateral hemiplegia: Evidence for viral invasion of cerebral arteries. Neurology. 1996;**47**(2):569-570

[21] Gilden DH, Kleinschmidt-DeMasters BK, Wellish M, Hedley-Whyte ET, Rentier B, Mahalingam R. Varicella zoster virus, a cause of waxing and waning vasculitis: The New England Journal of Medicine case 5-1995 revisited. Neurology. 1996;**47**(6):1441-1446

[22] Linnemann CC Jr, Alvira MM. Pathogenesis of varicella-zoster angiitis in the CNS. Archives of Neurology. 1980;**37**(4):239-240

[23] Sreenivasan N, Basit S, Wohlfahrt J, Pasternak B, Munch TN, Nielsen LP, et al. The short- and long-term risk of stroke after herpes zoster - a nationwide population-based cohort study. PLoS One. 2013;**8**(7):e69156. DOI: 10.1371/journal.pone.0069156

[24] Langan SM, Minassian C, Smeeth L, Thomas SL. Risk of stroke following herpes zoster: A self-controlled case-series study. Clinical Infectious Diseases. 2014;**58**:1497-1503. DOI: 10.1093/cid/ciu098

[25] Askalan R, Laughlin S, Mayank S, Chan A, MacGregor D, Andrew M, et al. Chickenpox and stroke in childhood: A study of frequency and causation. Stroke. 2001;**32**:1257-1262

[26] Mishell JH, Applebaum EL. Ramsay-Hunt syndrome in a patient with HIV infection. Otolaryngology and Head and Neck Surgery. 1990;**102**(2):177-179. DOI: 10.1177/019459989010200215

[27] Adour KK. Otological complications of herpes zoster. Annals of Neurology. 1994;**35**(Suppl):S62-S64

[28] Furuta Y, Takasu T, Fukuda S, Sato-Matsumura KC, Inuyama Y, Hondo R, et al. Detection of varicella-zoster virus DNA in human geniculate ganglia by polymerase chain reaction. The Journal of Infectious Diseases. 1992;**166**(5):1157-1159

[29] Young-Barbee C, Hall DA, LoPresti JJ, Schmid DS, Gilden DH. Brown-Séquard syndrome after herpes zoster. Neurology. 2009;**72**:670-671. DOI: 10.1212/01.wnl.0000343737.79883.5c

[30] Steiner I, Kennedy PG, Pachner AR. The neurotropic herpes viruses: Herpes simplex and varicella-zoster. Lancet Neurology. 2007;**6**(11):1015-1028. DOI: 10.1016/S1474-4422(07)70267-3

[31] Gilden DH, Beinlich BR, Rubinstien EM, Stommel E, Swenson R, Rubinstein D, et al. Varicella-zoster virus myelitis: An expanding spectrum. Neurology. 1994;**44**:1818-1823

[32] Rowbotham M, Harden N, Stacey B, Bernstein P, Magnus-Miller L. Gabapentin for the treatment of postherpetic neuralgia: A randomized controlled trial. Journal of the American Medical Association. 1998;**280**(21):1837-1842

[33] Choo PW, Galil K, Donahue JG, Walker AM, Spiegelman D, Platt R. Risk factors for postherpetic neuralgia. Archives of Internal Medicine. 1997;**157**(11):1217-1224

[34] Nalamachu S, Morley-Forster P. Diagnosing and managing postherpetic neuralgia. Drugs & Aging. 2012;**29**(11):863-869. DOI: 10.1007/s40266-012-0014-3

[35] Hempenstall K, Nurmikko TJ, Johnson RW, A'Hern RP, Rice AS. Analgesic therapy in postherpetic neuralgia: A quantitative systematic review. PLoS Medicine. 2005;**2**(7):e164. DOI: 10.1371/journal.pmed.0020164

[36] Edelsberg JS, Lord C, Oster G. Systematic review and meta-analysis of efficacy, safety, and tolerability data from randomized controlled trials of drugs used to treat postherpetic neuralgia. The Annals of Pharmacotherapy. 2011;**45**(12):1483-1490. DOI: 10.1345/aph.1P777

[37] Attal N, Cruccu G, Baron R, Haanpää M, Hansson P, Jensen TS, et al. European Federation of Neurological Societies. EFNS guidelines on the pharmacological treatment of neuropathic pain: 2010 revision. European Journal of Neurology. 2010;**17**(9):1113-e88. DOI: 10.1111/j.1468-1331.2010.02999.x

[38] Levin MJ, Oxman MN, Zhang JH, Johnson GR, Stanley H, Hayward AR, et al. Varicella-zoster virus-specific immune responses in elderly recipients of a herpes zoster vaccine. The Journal of Infectious Diseases. 2008;**197**(6):825-835. DOI: 10.1086/528696

[39] Oxman MN, Levin MJ, Johnson GR, Schmader KE, Straus SE, Gelb LD, et al. A vaccine to prevent herpes zoster and postherpetic neuralgia in older adults. The New England Journal of Medicine. 2005;**352**(22):2271-2284. DOI: 10.1056/NEJMoa051016

[40] Nagel MA, Cohrs RJ, Mahalingam R, Wellish MC, Forghani B, Schiller A, et al. The varicella zoster virus vasculopathies: Clinical, CSF, imaging, and virologic features. Neurology. 2008;**70**(11):853-860. DOI: 10.1212/01.wnl.0000304747.38502.e8

[41] Schubert J, Weissbrich B. Detection of virus-specific intrathecally synthesised immunoglobulin G with a fully automated enzyme immunoassay system. BMC Neurology. 2007;**7**:12. DOI: 10.1186/1471-2377-7-12

[42] Nagel MA, Forghani B, Mahalingam R, Wellish MC, Cohrs RJ, Russman AN, et al. The value of detecting anti-VZV IgG antibody in CSF to diagnose VZV vasculopathy. Neurology. 2008;**68**(13):1069-1073. DOI: 10.1212/01.wnl.0000258549.13334.16

[43] Solomon T, Michael BD, Smith PE, Sanderson F, Davies NW, Hart IJ, et al. Management of suspected viral encephalitis in adults: Association of British Neurologists and British Infection Association National Guidelines. The Journal of Infection. 2012;**64**:347-373. DOI: 10.1016/j.jinf.2011.11.014

Chapter 4

Extracranial Herpetic Paresis

Vesna Martic

Abstract

Segmental zoster paresis (SZP) is a rare complication of varicella zoster infection that occurs due to the spread of the infection from the posterior horn of spinal cord to the anterior horn and the motor nerve root. As recognizing segmental zoster paresis is important in the differential diagnosis of muscle weakness of other origin, information about demographic (gender and age), clinical presentation, diagnosis, treatment, and course about published patients with SZP was extracted from PubMed database. SZP is classified into several categories: paresis of upper extremity, lower limb involvement, diaphragmatic involvement, and abdomen involvement. Published experiences have shown that clinical course and electromyoneurography of paretic muscle are the most important in the diagnosis; physical therapy is the most common therapy in these patients and their prognosis is generally good except diaphragmatic paresis, where there is no significant recovery in most number of patients.

Keywords: segmental paresis, varicella zoster, clinical presentation, diagnosis, treatment, course

1. Introduction

Herpes zoster (HZ) is an infection of dorsal root ganglion characterized by a painful cutaneous rash. It is believed that reactivated varicella zoster virus (VZV) migrate from dorsal root ganglion in retrograde direction along the sensory nerve to the skin, where it makes the characteristic dermatomal rash [1].

The initial infection caused by varicella or chickenpox occurs during childhood. After that, the virus then exists as a latent infection of sensory ganglia from which it may reactivate many years later and cause herpes zoster.

The virus can be recovered from skin lesion and can cause an inflammatory reaction in sensory ganglion, dorsal root, and posterior horn of spinal cord [2].

Postherpetic neuralgia is the most common complication associated with extracranial HZ. Pain persisting 90 days or longer after the onset of the shingles rash is called postherpetic neuralgia [3, 4]. Postherpetic neuralgia usually remits spontaneously, but some patients could have pain for all their life.

Beside postherpetic neuralgia, segmental herpetic paresis is another complication of HZ, and they often occur together.

Flaccid muscle paralysis rarely occurs due to the spread of the infection from the posterior horn of spinal cord to the anterior horn and the motor nerve root (**Figure 1**) when weakness generally corresponding to the dermatomes in which cutaneous lesions develop [5].

This finding supports the enhancement of spinal nerve roots on MRI that was clinically symptomatic [6], attributed to autoimmune inflammation or

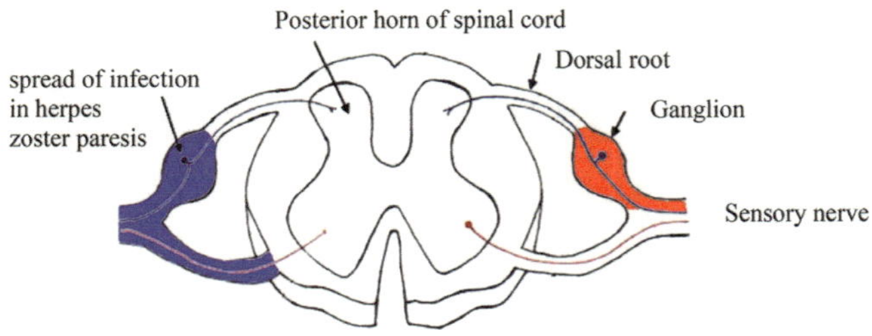

Figure 1.
Origin of segmental zoster paresis: the spread of varicella zoster infection from the posterior horn of spinal cord to the anterior horn and the motor nerve root.

vasculitis [7]; brachial plexus inflammation; and myelin destruction with intact axons found postmortem [8].

The association of muscle paralysis and herpes zoster was first reported by Broadbent in 1866. When he described the case to zoster described as "frozen shoulder" [9], it is noted rarely between 0.5% [10] and 0.8% of segmental motor paralysis between all patients with HZ [11].

In patients with dissociation between motor and dermatomal involvement, a possible explanation could be viral spread to anterior roots without corresponding axonal transport through the sensory nerves [6, 12].

Some patients may have zoster infection without vesicular eruptions, which is called "zoster sine herpete." This diagnose is safer by a rise VZV antibodies. These patients may have the same neurologic manifestations, including muscular paralysis [13].

As there are no certain standards regarding diagnosis and treatment of segment zoster paresis, we have listed in this paper experiences from literature related to the topic.

The aim of this study was to summarize the experiences related to its clinical manifestation, applied diagnostics, treatment, and patient outcome.

2. Methods

We searched the PubMed database for literature on herpes zoster infection and extracranial motor paresis in adults (last search on September 2019).The search was limited to full-length articles written in English and a study population that included adults aged 18 years and older. A combination of the following search terms was used "herpes zoster, paresis, complications." As a result, 74 articles were retrieved.

The reference list was also searched for relevant manuscripts not retrieved from PubMed.

Studies included in the final review met the following criteria: (1) infection with herpes zoster virus, (2) motor paresis, and (3) adult study population.

They encompass original articles, technical reports, clinical observations, and single case reports.

3. Results

3.1 Data extraction and synthesis

For comparison across the reports, the sample size was extracted along with demographic information (gender and age), clinical presentation, laboratory confirmation of herpes zoster, electromyographic (EMG) findings, imagining, treatment, and course.

For clarity, segmental zoster paresis is classified into several categories: paresis of upper extremity, lower limb involvement, diaphragmatic involvement, and abdomen involvement.

Upper extremity involvement: In 37 previously published papers about patients with segmental zoster paresis of arms, there are 19 papers that are processed only as arms paresis (**Table 1**), while the rest of the 18 papers describe patients with segmental zoster paresis of arms and legs (**Table 2**). They include a total of 101 patients with segmental paresis of arms with a mean age of 68.56 ± 11.97 and with 55 women and 45 men and one patient who did not specify gender.

There are few comorbidities in this population of patients: 2 patients with carcinoma mammae, 7 patients suffering from lymphoma, 4 were taking corticosteroids due to autoimmune diseases (2 patients with rheumatoid arthritis, 1 with polymyalgia rheumatica, and 1 with myasthenia gravis), and 4 of them were diabetic.

Clinically, segmental zoster paresis is usually unilateral paresis of arm, predominantly on the right side (42/26, 33 cases do not specify the affected side).

Proximal muscles were affected in most cases (in 52 patients), while the entire arm [40] or only distal muscles [41] were significantly less affected.

Besides weakness, in 9 patients, muscle atrophy was detected.

Phrenic nerve affection followed by dyspnea was observed in 6 patients.

Electromyoneurography was performed in 58 patients, and denervation potential has been described in most cases (51 patients), often associated with reduced pattern and polyphasia of motor unit potentials in most cases (17 patients). Reduced motor or sensor velocity was seen in 4 patients. In some cases, neurophysiological finding was only descriptive: "lesion of nerve," "acute motor axonal lesion," and "severe median and ulnar nerve neuropathy."

Another test was performed in only small number of patients: neck MRI (unremarkable in 2 patients, foraminal stenosis in 1, and protruded disk in 1) and MRI of the affected arm in 1 patient with enlargement of T2 signal of the median nerve.

X-ray of the shoulder was performed in 2 patients (1 patient showed subluxation) and 2 X-rays of the neck (1 unremarkable and 1 with spondyloarthrosis) and 2 X-rays of the chest with 1 showing elevated hemidiaphragm and paralysis confirmed by following fluoroscopy were performed.

Standard laboratory in 2 patients was unremarkable; positive sera antibody test for HZ was found in 3 patients.

Hyperproteinorachia was found in 2 patients and normal CSF was found in 1 patient among patients with lumbar puncture.

In 1 patient, surgical exploration was done and it was without compression.

In almost all cases with a mentioned type of treatment were treated with physical therapy (24 reported patients). Some of them had other therapies: Valacyclovir 3 g/7 days (3 patients), Acyclovir 750 mg/7 days (2 patients iv and 1 per os); and Methylprednisolone 500 mg iv for 3 days (1 patient), and some of them were taking steroids [11], cervical epidural bloc [11], analgesics [42], opioids [43], Amitriptyline [43], Pregabalin [11], and Gabapentin [11].

Most reported patients recovered significantly: complete or near complete recovering is recorded in 8 patients that are reported during following periods: 3 weeks, 1 month, 2 months, 3 months, 6 months, and 2 years. Incomplete recovery was reported in 3 patients, minimal recovery in 1 patient after 2 months, and no recovery after 8 months in 1 patient. One patient recovered from weakness after 5 months but with persisted neuralgia.

Literature	Demographics, medical history	Subjective complaints	Neurological impairments	EMG findings	Other tests	Treatment	Course/Outcome
[14]	58-year old. female	Rash in left shoulder, lateral part of left arm and hand; pain in neck and interscapular region	Weakness of left shoulder, wasting of left *M. deltoideus*			Exercises	After 2 months, power and mass of M. deltoideus increase
[15]	#1: 82-year old female	Pain and rash in right shoulder and upper arm	Weakness in proximal arm mild elbow flexion contracture	Denervation in deltoid, biceps, supraspinatus; polyphasia and decreased number of MUAP		Moist hot packs, ultrasound, exercise	Incomplete recovery
	#2: 72-year old female, breast carcinoma with mastectomy		Weakness in biceps, deltoid, wrist flexors/extensors	Denervation in almost all arm muscles; ulnar and median conduction velocities reduced	Not reported	B12 injections, heat, paraffin baths, exercises	Incomplete recovery
	#3: 65-year old female with knee amputation for vascular disease, chronic leukemia	Pain and rash in right arm	Weakness in deltoid, biceps, triceps, wrist extensors, thenar muscles	Marked denervation	Not reported	Heat, electrical stimulation, strengthening exercises	Functional recovery within 6 weeks
	#4: 62-year old female, rheumatoid arthritis	Pain and rash in right arm	Weakness in deltoid, infraspinatus, supraspinatus, biceps	Denervation; polyphasic MUAP	Not reported	Heat, electrical stimulationsling	Some recovery of strength and less pain
	#5: 71-year old female	Pain and rash in right shoulder and forearm	Weakness in deltoid, biceps, wrist flexors/extensors	Denervation in C7–8/T1 muscles	Not reported	Codeine, whirlpool, paraffin baths	Nearly complete recovery after unknown time period
[16]	#1: 70-year old female	Pain in left shoulder with radiation to arm, hyperesthesia in shoulder and neck; rash in C4–5 dermatomes	Weakness in arm; reflexes absent in biceps, supinator	Denervation in deltoid and biceps muscles	Unremarkable chest X-ray and laboratory work-up	Physical therapy	Full recovery after 3 weeks

DOI: http://dx.doi.org/10.5772/intechopen.90493

Literature	Demographics, medical history	Subjective complaints	Neurological impairments	EMG findings	Other tests	Treatment	Course/Outcome
	#2. 60-year old man, lymphatic leukemia	Rash and hypersensitivity in right C5 dermatome after weeks of coughing; enlarged liver		Not reported	Chest X-ray: elevated right hemidiaphragm, paralysis confirmed on fluoroscopy		Complete recovery after 1 month
[17]	25-year old female	Rash over shoulder	Weakness in deltoid	Not reported	X-ray: shoulder subluxation	Not reported	Full recovery after unspecified number of months
[18]	59-year old man, diabetes, mild hypertension	Pain and rash in left C7 dermatome; hyperalgesia in inner half of left hand	Weakness in whole arm; marked atrophy in almost all arm muscles, reflexes diminished	Fibrillations, reduced interference pattern in atrophied muscles; conduction velocities normal	X-ray: spine normal; CSF: normal	Not reported	Unknown
[13]	#1. 81-year old man	Pain in left arm and left lateral chest wall; rash in C5–7 dermatomes	Complete flaccid weakness of left arm	Not reported	Not reported	Physical therapy	Complete recovery after 3 months
	#2. 75-year old female, diabetes	Pain and rash in right shoulder and over back of the head	Marked weakness in deltoid, biceps, infraspinatus; reflexes absent in biceps	Denervation in infraspinatus, deltoid, biceps; conduction velocities normal	Not reported	Not reported	Minimal recovery after 2 months
[5]	64-year old female	Pain and rash in left inner arm and fingers after 4 months of an episode of shingles	Severe weakness in wrist and hand intrinsic muscles	Lesion of anterior interosseus nerve	Surgical exploration-no compression	Not reported	No recovery after 8 months
[19]	73-year old man, left biceps rupture 10 yrs. earlier with full recovery	Pain in left shoulder; rash over biceps, brachioradialis	Moderate atrophy of deltoid; weakness in biceps and infraspinatus; reflexes	Acute motor axonal lesion in C5–6 dermatomes	Neck MRI: no spinal cord, root compression	Acyclovir, opioid analgesics, amitriptyline, physical therapy, home exercises	Slight deltoid weakness after 2 years

Literature	Demographics, medical history	Subjective complaints	Neurological impairments	EMG findings	Other tests	Treatment	Course/Outcome
			diminished in biceps, brachioradialis muscle				
[7]	72-year old man with oral prednisolone due to myasthenia gravis	Rash in left C3–5 dermatomes	10 days before the rash, weakness of the left arm limited to muscles controlled by the C5 myeloma		T2-weighted MRI		
[20]	73-year old man	Right shoulder pain and herpes zoster eruptions over the C5 dermatome	Two days later, he found it impossible to lift up the right arm				
[21]	A 48-year old female	Rash and vesicles over her left C5–7 dermatomes	Weakness in the left deltoids and biceps muscles, and a diminished left biceps reflex; dyspnea with paradoxical abdominal wall movement		Chest CT normal, MRI-hyperintensity in the left anterolateral aspect of the spinal cord at C5 level	1000 mg valacyclovir orally three times daily for 7 days.	The patient's pain resolved three months later, and the patient underwent outpatient follow-up for 2 years without further complications
[22]	A 72-year old female	Burning pain and rash in the left shoulder and its weakness	Hyperalgesia skin in left C4–T2 dermatomes, left shoulder abduction weakness	EMG-denervation in left biceps, deltoid, brachioradialis and C5–6 paraspinal muscles; sensory and motor nerve-conduction, and somatosensory evoked potentials are normal	MRI of cervical spine-degenerative changes with foraminal stenosis at C3–T1	750 mg/d famciclovir per os, cervical epidural blocks with 20 mg triamcinolone twice during 4 weeks, 450 mg pregabalin and 10 mg nortriptyline, physiotherapy	After 2 months, pain and weakness were significantly retracted
[23]	88-year old man with corticosteroids in therapy because of polymyalgia rheumatica	Rash in the right C3–5 dermatome	Paresis in the right shoulder 2 days after the rash			Oral valaciclovir, acyclovir iv after paresis and methylprednisolone	Paresis has gradually improved

Literature	Demographics, medical history	Subjective complaints	Neurological impairments	EMG findings	Other tests	Treatment	Course/Outcome
[24]	59-year old-female with carcinoma mammae dextri and lung metastasis, chemotherapy and radiotherapy	Pain and rash in whole right arm and scapula	Patient was able to adduct shoulder, flexor and extensor wrist and passively extend elbow	EMNG-low amplitude action potentials in left superior radial and median, SNAPs and MUAPs in left deltoid muscle	Positive direct fluorescent antibody test for VZV, neck MRI-C4-C5-C6 levels protruded disk	Acyclovir 10 mg/ 8 hours i.v. for 7 days	Motor activities recovered after 5 months with physiotherapy but postherpetic neuritis had a poor decrement

Table 1.
List of studies that reviewed herpes zoster infection and motor paresis of only upper limbs.

Literature	Demographics, medical history	Subjective complaints	Neurological impairments	EMG findings	Other tests	Treatment	Course/outcome
[25]	#1: 89-year old man, transient facial weakness 1 year earlier, conjunctivitis 6 months earlier	Pain and rash in right thigh and knee	Weakness and atrophy in right thigh, decreased sensation; reflexes diminished in knees and ankles	Denervation in *M. quadriceps*, tibialis anterior, peroneus longus and gastrocnemius	X-ray: degenerative changes in LS spine	Physical therapy, galvanic electrical stimulation	Incomplete recovery after 3 months
	#2. 44-year old female	Pain and rash over right distal arm in C7–8 dermatome	Weakness of proximal and distal muscles of right arm	Fibrillations in right upper arm in C8 myotome	Not reported	Physical therapy	Incomplete recovery
	#3. 74-year old woman	Pain and rash in T12-L1 dermatomes	Bulging of right abdomen and decreased muscle tone	Not reported	Not reported	Corset	Not reported
	#4. 58-year old woman, diabetes	Pain in left leg and rash over S1–2 dermatomes	Weakness in left knee extensors, ankle dorsiflexors and plantar flexors; reflexes absent in left leg	Not reported	Not reported	Not reported	Able to walk after 5 weeks
[26]	#1. 71-year old female	Rash along right ulnar border of forearm and hand	Weakness in forearm and hand muscles	Not reported	Not reported	Physical therapy	Return of strength but impaired fine movements
	#2. 76-year old female	Rash over right shoulder and arm	Weakness in complete right arm, reflex absent in biceps	Not reported	Not reported	Physical therapy	Full recovery except in deltoid muscle, neuralgia over 3 years
	#3. 83-year old female	Pain and rash over right shoulder	Weakness in deltoid, biceps; reflexes absent in biceps	Not reported	Not reported	Hydrotherapy, exercises	Weakness in deltoid and biceps; neuralgia over 8 months
	#4. 85-year old woman	Pain and rash in left shoulder	Weakness in deltoid, supraspinatus biceps; reflex diminished in triceps	Not reported	Not reported	Physical therapy	Incomplete recovery, residual neuralgia died after 7 months

DOI: http://dx.doi.org/10.5772/intechopen.90493

Literature	Demographics, medical history	Subjective complaints	Neurological impairments	EMG findings	Other tests	Treatment	Course/outcome
	#5. 73-year old man	Pain and rash in right shoulder and in upper arm	Weakness in shoulder with deltoid fasciculations	Not reported	Not reported	Intensive physical therapy	Full recovery after 7 months
	#6. 74-year old woman	Rash over left thigh	Weakness of hip flexors, adductors, knee extensors	Not reported	Not reported	Not reported	Full recovery after 4 months
	#7. 73-year old man	Rash over left shoulder and upper third of arm	Weakness in left arm	Not reported	Not reported	Hydrotherapy, physical therapy	Full recovery
	#8. 81-year old female	Pain and rash over C1–7 dermatomes	Weakness of right serratus anterior, scapula winging	Not reported	Not reported	Intensive physiotherapy	Full recovery in arm, residual pain after 5 months
	#9. 75-year old female	Rash of outer part of right arm	Weakness in shoulder flexors and abductors, and in elbow flexors	Not reported	Not reported	Intensive physiotherapy	Full recovery after 5 months; some pain persisted
[27]	#1. 73-year old female	Rash in right C4–7 dermatomes; segmental severe paresis in C6, 7, 8		Denervation	CSF normal; myelogram normal	Supportive treatment	Full recovery after 7 months
	#2. 78-year old female, diabetes	Rash in right C5–7 dermatomes	Moderate weakness in C5–7 myotomes	Denervation	CSF normal; myelogram normal	Supportive treatment	Full recovery after 9 months
	#3. 67-year old man	Rash in left C5–6 dermatomes	Moderate weakness in C5–6 myotomes	Denervation	Not reported	Supportive treatment	Full recovery after 3 months
[28]	#1. 84-year old man	Pain and rash in right L2–3 dermatomes; impaired sensation in L2–3 dermatomes	Weakness in quadriceps and hip adductors, and flexors; reflexes absent in knee	Fibrillations in hip flexors, vastus medialis	Not reported	Not reported	No recovery, died 6 years later

Literature	Demographics, medical history	Subjective complaints	Neurological impairments	EMG findings	Other tests	Treatment	Course/outcome
	# 2. 66-year old man	Rash and pain in left shoulder and arm in C5 dermatome	Weakness in deltoid and spinatus muscles	Fibrillation, polyphasia and reduced interference patterns MUAP	Not reported	Not reported	Complete recovery after 4 months
	#3. 85-year old female	Rash, pain and impaired sensation in left in C4–5 dermatomes	Weakness in deltoid, spinatus, biceps; reflexes absent	Fibrillations and reduced interference pattern in deltoid	Not reported	Not reported	Minimal weakness after 4 years, residual neuralgia
	#4. 83-year old. female	Pain in outer part of right arm and digitis III-V, rash on posterior forearm and digitis	Weakness in biceps, in triceps, fingers, hand intrinsics; absent reflex in triceps, depressed in biceps, supinator	Fibrillations polyphasia and reduced interference patterns, motor median velocity reduced; absent median sensory neurogram	Not reported	Not reported	Incomplete recovery of hand after 1 year, full recovery in triceps wrist extensors
	#5. 64-year old man	Pain in right shoulder, rash in C5 distribution	Weakness in deltoideus, supraspinatus; impaired sensation in C5 dermatome; reflexes absent in whole arm	Fibrillations in deltoid, moderately reduced interference patterns in deltoid and supraspinatus	Not reported	Not reported	Full motor recovery after 2 years
	C#6. 52-year old female	Pain and rash in left arm (C5–6 distribution); numbness in left thumb	Weakness in left deltoid, spinatus, biceps; reflexes absent in biceps, supinator	Reduced interference pattern in left deltoideus with polyphasic units	Not reported	Not reported	Incomplete recovery after 7 months
	#7. 77-year old man	Pain in right groin, rash in anterior and medial thigh	Weakness in hip flexors, adductors, knee extensors; reflex absent at knee	Fibrillations in thigh muscles, reduced patterns, polyphasic units	Not reported	Not reported	Full recovery after 4 months

Literature	Demographics, medical history	Subjective complaints	Neurological impairments	EMG findings	Other tests	Treatment	Course/outcome
	#8. 73-year old man	Pain around knee, rash in medial aspect of left thigh	Wasting and moderate weakness in quadriceps; reflex diminished at the knee	Fibrillations slightly reduced interference pattern, polyphasic units	Not reported	Not reported	Full recovery after 4 months
	#9.69-year old man	Pain in right axilla, rash in medial arm and medial aspects of two fingers	Weakness in hand intrinstics, sensory impairment in C8, T1 and T2 dermatomes; absent reflexes	Fibrillations polyphasia and reduced interference pattern, reduced motor velocity in ulnar nerve, absent sensory neurograms	Not reported	Not reported	Marked wasting and weakness in thenar, hypothenar, and intrinsic muscles after 1 year
[29]	#1. 77-year old man	Pain and rash in left shoulder	Weakness in shoulder, biceps, triceps, and wrist extensors muscles	Denervation in deltoid and biceps, reduced MUAP recruitment	Routine lab normal; CSF increased protein, normal cell count; VZV antibodies IgG positive IgM negative; cervical spine MRI normal	Valacyclovir 3 g for 7 days, Acyclovir iv 750 mg for 7 days, Methylprednisolone iv 500 mg for 3 days, Prednisol 60 mg and tapering	Full recovery after 1 year
	#2. 57-year old man with diabetes	Pain and rash over right wrist and groin unable to walk	Weakness in right hip, thigh muscles, ankle dorsiflexors, patellar reflex diminished	Denervation in iliopsoas, quadriceps, tibial, F waves reduced in frequency	Blood tests normal; CSF increased protein and cell count; VZV antibodies IgG and IgM positive; lumbar spine MRI normal	Acyclovir 750 mg for 7 days, Acyclovir iv 750 mg for 7 days, Methylprednisolone iv 1000 mg for 3 days, Prednisolone 60 mg	Full recovery after 3 months
	#3. 65-year old female	Pain and rash in left shoulder and arm C5–6 dermatomes	Weakness in shoulder, reflexes diminished	Reduced interference with denervation in deltoid and C5 paraspinals muscles	Cervical spine MRI normal	Valacyclovir 3 g	Marked recovery after 3 months
[11]	#1.84-year old man with myelodysplastic syndrome	R C5,6	Weakness in C5,6 and atrophia	Denervation, polyphasia, normal conduction velocity			Moderate recovery after 6 months

Literature	Demographics, medical history	Subjective complaints	Neurological impairments	EMG findings	Other tests	Treatment	Course/outcome
	#2. 85-year old female	Pain and rash in left C5	Weakness and atrophia in left C5,6 distribution	Denervation, polyphasia, normal conduction velocity			Poor recovery after 1 year and 7 months
	#3. 79-year old female	Pain and rash in right C 5,6	Weakness in 5, 6 distribution	Denervation, polyphasia			Moderate recovery after 2 months
	#4. 67-year old female	Pain and rash in right C5,7 dermatome	Weakness in C5–8	Denervation, polyphasia			Good outcome after 6 y + 8 m
	#5. 82/F	Pain and rash in dermatome L1-S2 bilateral	Weakness in projection L1-S1 bilateral				Moderate recovery after 3y + 8 m
	#6. 80-year old man	Pain and rash in left C6,7	Weakness in left C5-Th1				Good recovery after 1y + 8 m
	#7. 76-year old female	Pain and rash in right C6,7 dermatome	Weakness in C6-Th1 distribution				Good recovery after 7y
	#8. 83-year old male	Pain and rash in right C8-Th1 dermatome	Weakness in C7, 8 distribution				Good recovery after 5y + 9 m
	#9. 7 year old female	Pain and rash in right L4-S1 dermatome	Weakness in right L1, S1 distribution				Good recovery after 5y + 4 m
	#10. 72-year old female	Pain and rash in right C5–7 dermatome	Weakness in right C5-Th1 distribution				Moderate recovery after 2 years and 10 month
	#11. 43-year old male bone marrow transplantation	Rash and pain in right C8-Th1 dermatome	Weakness in C5-Th1 distribution				Uncertain 6y + 8 m
[30]	#1. 69-year old female with diabetes	Pain and rash in right arm, rash over face	Weakness and impaired sensation in shoulder, reflexes diminished in right arm	Not reported	Not reported	Steroid, procaine, physical therapy	Almost full recovery

Literature	Demographics, medical history	Subjective complaints	Neurological impairments	EMG findings	Other tests	Treatment	Course/outcome
	#2. 67-year old man	Pain and rash in low back and right leg in distribution of the sciatic nerve	Weakness in right hip flexors; knee jerk absent	Not reported	Not reported	Paravertebral sympathetic block	Unknown
[31]	#1. 63-year old man	Pain and rash in right shoulder, arm, and hand	Weakness in shoulder, elbow flexors, extensors, in hand muscles; reflexes absent	Denervation in hand, motor amplitudes decreased, velocity normal; absent sensory neurograms	Complement fixing for antibodies VZV in sera was elevated	Physical therapy	Significant recovery except hand intrinsics
	#2. 80-year old woman	Pain from left knee to foot, rash in L5-S2 dermatomes, urinary frequency and incontinence	Weakness in knee flexors, ankle flexors and extensors; reflexes in ankles absent	Not reported	CSF: hyperproteinorachia	Not reported	Incomplete recovery: foot drop and urinary retention remained
[32]	15 patients (9 females and 6 males, mean age 66 years, range 48–80) with rheumatoid arthritis in 3, lymphosarcoma in 1, and lymphatic leukemia in 1	Pain-to-rash interval 2–3 days in 5 patients, 4–6 days in 6, 7 days in 2, unknown in 2; Rash-to-weakness interval < 10 days in 1 patient, 10 to 28 days in 12, 42 days in 1, unknown in 1 (similar in upper and lower limbs)	Distribution of motor paresis: C5–6 in 5, C5–7 in 1, C7-C8-T1 in 2; L3-S1 in 7, L3-S1. Weakness severe in 10 patients, moderate in 3, mild in 2; sensory abnormalities in half of the patients; reflexes diminished	Denervation in 12 patients	Increased protein and cell count in CSF of one patient	Not reported	Full recovery in 11 patients (5 arms and 6 legs). Mean recovery time in arms 9 months, 7 months in legs. 2 improved arms and 1 in legs. Postherpetic neuralgia in 3 patients
	#1. 70-year old man	Pain and rash in C5 myotome	Not reported	Not reported	Not reported	Not reported	Marked recovery
	#2. 56-year old man	Pain and rash in L3 myotome	Diminished knee reflexes	Not reported	Not reported	Not reported	Marked recovery
	#3. 59-year old man	Pain and rash in C5 myotome	Absent reflexes	Not reported	Not reported	Not reported	Full recovery

Literature	Demographics, medical history	Subjective complaints	Neurological impairments	EMG findings	Other tests	Treatment	Course/outcome
	#4. 70-year old man	Pain and rash in C5 myotome	Absent reflexes	Not reported	Not reported	Not reported	Marked recovery
	#5. 71-year old man	Pain and rash in C5–6 myotome	Absent reflexes	Not reported	Not reported	Not reported	Full recovery
	#6. 67-year old female	Pain and rash in C7 myotome	Not reported	Not reported	Not reported	Not reported	Full recovery
	#7. 91-year old woman	Pain and rash in T1 myotome	Not reported	Not reported	Not reported	Not reported	No recovery
	#8. 65-year old female	Pain and rash in C5 myotome	Absent reflexes	Not reported	Not reported	Not reported	Full recovery
	#9. 72-year old man	Pain and rash in C5 myotome	Absent reflexes	Not reported	Not reported	Not reported	Unknown
	#10. 65-year old man	Pain and rash in L2–3 myotome	Absent knee reflexes	Not reported	Not reported	Not reported	Marked recovery
	#11. 56-year old man	Pain and rash in L5 myotome	Not reported	Not reported	Not reported	Not reported	No recovery
	#12. 76-year old man	Pain and rash in C5 myotome	Absent SJ and BJ	Not reported	Not reported	Not reported	Modest recovery
	#13. 62-year old woman	Pain and rash in L3 myotome	Absent knee reflex	Not reported	Not reported	Not reported	Modest recovery
	#14. 70-year old woman	Pain and rash in L3 myotome	Absent knee reflex	Not reported	Not reported	Not reported	Full recovery
[33]	61 patients (39 men and 22 women), mean age 62 yrs. (range 18–87); lymphoma in 6, chronic lymphocytic	Rash-to-weakness interval in 51 patients: <3 days in 6, 3–6 days in 12, 7–10 days in 14, 11–14 days in 11, 15–	Weakness in upper limbs in 16 (C5 to T1 segments at about equal frequency), lower limbs in 15	EMG in 18 patients (9 of them are within legs): fibrillations present in all affected muscles, MUAPs	CSF proteins and cell count increased in 2 patients	Not reported	Limb recovery full in 55%, marked in 25%; residual pain in 8 patients (4 of them

Literature	Demographics, medical history	Subjective complaints	Neurological impairments	EMG findings	Other tests	Treatment	Course/outcome
	leukemia in 1, diabetes in 3, histoplasmosis in 1	20 days in 5, 21–28 days in 2 patients, 29–35 days in 1	patients (L2 to S1 at about similar frequency); abdominal weakness in 2 patients	decreased in number in 2 with large amplitudes; sensory and motor nerve conduction studies normal in all but 3 cases			are with leg distribution)
[34]	#1. 45-year old woman	Pain in right thigh, rash along the sciatic nerve	Decreased sensation in L2-S1-S2 dermatomes, reflexes diminished; weakness in right ankle dorsi/plantar flexors	7 years later- high amplitude and polyphasia on MUAP	Lumbar/pelvic X-ray and CSF normal	Not reported	Full recovery at 3 months after the first episode; moderate after each of next relapses over 7 years
	#2. 80-year old man	Pain in all right arm, rash over anterolateral part of arm	Weakness in all right arm, reflexes diminished, shoulder atrophy and subluxation of humeral head	Fasciculations, rare action potentials in deltoid, supraspinatus and biceps muscles	Neck and spine X-ray: spondyloarthritic changes	Not reported	Incomplete recovery in shoulder muscles after 1.5 years
[35]	#1. 77-year old woman with hypertension and cardiomegaly	Pain and rash in right thigh, decreased sensation in L1–3 dermatomes	Weakness and atrophy in quadriceps; knee reflex absent	Not reported	Not reported	Analgesics, physical therapy	Full recovery after 3–4 months
	#2. 65-year old man	Pain in left chest radiating in left arm, rash in left C5–6 dermatomes	Weakness in proximal and distal muscles of left arm; reflexes absent in arm	Not reported	Not reported	Symptomatic treatment	Full recovery
	#3. 74-year old female with osteoarthritis	Pain and rash in right shoulder (C5–7)	Weakness in proximal and distal muscles of right arm	Denervation in upper and middle trunks of the brachial plexus	Chest X-ray normal	Not reported	Full recovery after 6 months

Literature	Demographics, medical history	Subjective complaints	Neurological impairments	EMG findings	Other tests	Treatment	Course/outcome
	#4. 67-year old female	Pain and rash in distal right arm, sensory loss in C6–8/T1 segments	Weakness and atrophy in hand muscles, reflexes diminished	Severe median and ulnar neuropathy	CSF increased protein, no cells	Analgesics, physical therapy	Full recovery after 8 months
	#5. 64-year old man	Rash over lower lateral chest on left side, T9–10 dermatome	Weakness in left rectus abdominis and oblique muscles	Denervation in external oblique muscles	Not reported	Mild analgesia	
	#6. 80-year old man with DM, myocardial infarction, CVI	Rash and burning pain over right lower abdomen	Bulging of lateral and anterior abdominal wall	Fibrillations and positive waves in right abdominal muscles with later reinnervation	Not reported	Not reported	Full recovery after 4 months
[14]	58-year old female	Pain over left side of neck, rash over left shoulder, lateral left arm, hand and wrist	Weakness of left shoulder and left deltoid muscle			Splint	After 2 months mass and power of the deltoid had increased
[36]	#1. 53-year old man	Disseminated vesicular rash and general malaise	Developed leg weakness in 2 days unable to stand; facial diplegia; reflexes depressed	Normal in arms and legs		Diagnosis of GBS was made and no specific treatment was given	One year later was totally asymptomatic
	#2. 69-year old man with asthma	Painful rash on right buttock	Bilateral facial and truncal weakness; weakness in legs all reflexes absent, loss of light touch and position sense		Mild cyanosis, reduced lung capacity	Received course of plasmapheresis	5 months later only residual sign of mild reduction of hip power
[21]	48-year old woman	Pain in left arm, rash over the C5 to C7 dermatomes	Weakness in the left deltoids and biceps muscle and a diminished left biceps reflex; dyspnea			1000 mg Valacyclovir orally three times daily for 7 days	After 2 years- without further complications; pain resolved in three months

Literature	Demographics, medical history	Subjective complaints	Neurological impairments	EMG findings	Other tests	Treatment	Course/outcome
[1]	80-year old female, dementia	Pain in left part of neck	Left facial palsy; difficulty walking with left lower limb				Gait was normal after 3 months, but left facial palsy remained complete
[37]	#1. 61-year old man	Painful rash on dorsum of right foot, sensory loss over the lateral right leg	Weakness of all parts of right leg, reduced right ankle reflex	Fibrillation and reduced MUAP in muscles innervated by the distal sciatic nerve	PCR of cutaneous crusted of right foot lesions was positive for VZV		No neurologic follow-up
	#2. 69-year old man	Burning pain in right upper limb	Weakness in muscles innervated by right median nerve (right hand)	Fibrillation potentials and reduced MUAP	MRI of arm enlargement T2 signal within the median nerve with gadolinium contrast	Gabapentin	Incomplete improvement-weakness with residual pain
	#3 83-year old woman	Rash over right upper limb	Weakness of right hand grip	Fibrillation and reduced MUAP in distribution of C7-T1 roots; conduction block in the median nerve			After 11 months moderate residual weakness in median innervated muscles
	#4 55-year old man with migraine, restless legs syndrome	Rash on knee and ankle, pain in left buttock, anterolateral thigh and knee	Left knee stretch reflex absent	Fibrillation and reduced MUAP in left iliopsoas and rectus femoris muscles	MRI enlargement and T2 signal in left femoral nerve		No clinical follow-up
[38]	60-year old man	Pain in right leg, rash in anteromedial part of right thigh	Weakness, atrophy and fasciculation of right quadriceps; right knee reflex absent	Reaction of degeneration in the right quadriceps femoris		Thiamine hydrochloride 10 mg. three times a day orally, heat and electrical stimulation	After 3 months incomplete recovery with atrophy of the thigh and fasciculations
[39]	#1. 71-year old man	Pain on right side of chest weakness in both leg and right hand	Deep reflexes brisk			Physiotherapy	3 months later almost full recovery

Literature	Demographics, medical history	Subjective complaints	Neurological impairments	EMG findings	Other tests	Treatment	Course/outcome
	#2. 58-year old man	Rash left side of neck and right upper arm	Weakness in both legs and right arm; reflexes absent in all limbs	Fibrillation potentials affecting muscles of right limb	Velocities affecting in right median and popliteal nerves	Physiotherapy, tetracycline	After 11 months little residual deficit
[10]	#1. 47-year old man	Rash in right shoulder and anterolateral arm	Weakness in right C5–6 myotomes	Denervation in C5–6 distribution			Recovery in 3 months
	#2. 70-year old female with DM	Rash in right lateral arm and forearm	After 20 days weakness in right C5–7 myotomes	Right brachial plexopathy with denervation	Hyperintensity in spinal dorsal horns at C4–5		No recovery after 2.0 years
	#3. 63-year old male with DM	Rash of the right foot and a right L5-S1 plexopathy	Weakness in right L5-S1 myotomes				No recovery after 1.8 years
	#4. 80–90-year old with DM	Rash in neck first	After 22 days right C8 myotome weakness	Denervation in right plexus brachialis			No recovery after 1.9 years
	#5. 87-year old male with DM	Rash in right lateral arm and forearm	After 14 days right C6–8 myotome weakness	Denervation in right C6–8 distribution	Increased signal in the C6–8 nerve roots		No recovery 1.0 year
	#6. 60-70-year (nn gender)	Rash in right buttocks and lateral calf	Weakness in right L5 myotome	Right L5 radiculoplexopathy with denervation			Partial recovery after 1.0 year
	#7. 61-year old male	Rash in left thumb, index finger and forearm	After 15 days weakness in left C6–8 myotomes	Left C7 radiculopathy with denervation	Increased signal in median and radial nerve on MRI		No recovery after 0.5 year
	#8. 80-year old female	Rash in right shoulder, anterolateral arm and thumb	Weakness in right C5 myotomes	Denervation in right C5 distribution	Increased signal in the C5 nerve roots on MRI		Partial recovery after 0.5 year

Table 2.
Lists of studies that reviewed herpes zoster infection and motor paresis of arms and legs.

3.1.1 Lower limb involvement

There are 43 presented patients with isolated segmental zoster paresis of one leg in the total of 26 previously published papers. Nine of these papers present only zoster paresis of the lower limbs, while the rest of them describe affection of arms and torso, also. According to available information, mean age of this group of patients was 64.19 ± 15.28, and the patient group was dominated by males (20/12).

From medical history, these patients had: diabetes mellitus (4 patients), lymphatic leukemia [11], ulcerative colitis [11], myelofibrosis [11], renal failure [11], hypertension arterialis [43], dementia [11], restless legs syndrome, and chronic low back pain [11].

Almost all patients had weakness in one leg; but 3 of them developed weakness in both legs.

In most cases, proximal muscles were affected (in 13 patients); then in much lesser number, distal muscles (6 patients) were affected; and only 4 of them had developed entire limb weakness.

Muscle atrophy was seen in 3 patients in this group (in *M. quadriceps*).

With regard to other complications, 2 patients developed incontinency, 1 urinary retention, 1 ileus, and 2 abdominal wall weakness.

Denervation has been the most common report (in 18 patients) among patients with electromyoneurography performed; reduced interference pattern was found in 12 patients and polyphasia was present in 4 patients.

Laboratory is sporadically performed in these patients: ELISA for HZV in sera done in 2 patients and were positive in IgG and IgM fraction in both of them, and in the third patient, performed VZV complement fixation was positive, also.

A lumbar puncture was done in a small number of patients [44, 45]: cerebrospinal fluid (CSF) was normal in two cases, but in the other three patients, hyperproteinorachia was discovered.

The patient with flaccid paraparesis had elevated lymphocytes and hyperproteinorachia in cerebrospinal fluid, while PCR for VZV was positive in this case.

MRI of lumbar spine performed in 4 patients did not explain the nature of deficit in these patients, and 2 patients were generally described by degenerative changes, 1 was normal, and in 1 enlargement and T2 signal in the left femoral nerve was found.

The most frequent treatments were physical therapy (4 patients), Acyclovir (2 patients), Methylprednisolone (2 patients), vitamins B1 and B12 [11], paravertebral sympathetic block [11], analgesics, and Gabapentin, lately [11].

There is no information about outcome of the disease for some patients. However, among patients who were followed for a period of time, a majority of them (20 patients) fully or almost fully recovered in the period between 2 and 9 months; 11 patients recovered incompletely; 2 patients did not recover, and 2 patients had with lethal outcome.

3.1.2 Diaphragmatic paralysis

Diaphragmatic paralysis due to phrenic nerve involvement with VZV is described in 24 previously published papers with a total of 26 patients with a mean age of 67.13 ± 11.50 years.

Among comorbidities in this population of patients are rheumatoid arthritis [11], leukemia [11], breast cancer [11], pyelonephritis and nephrectomy, hysterectomy [11], DM [11], hypertension [43], peptic ulcer [11], pancreatitis [11], and bypass [11].

Hemidiaphragmatic paresis is unilateral usually, predominantly on the left side [46], while in one patient, hemidiaphragmatic affection on both sides were recorded.

Although EMNG of phrenic nerve is one of the reliable signs of affection in this nerve, it is rarely performed (1 patient). More often, fluoroscopy is performed when weakness of the diaphragm is suspected (7 patients). Although not a reliable sign of weakness of the diaphragm, X-ray is often done in the case of a suspected weakness of the diaphragm paresis (13 patients).

Phrenic nerve affection is usually followed by dyspnea, but there is a case with hemidiaphragmatic paresis on X-ray, but without visible dyspnea.

To the contrary, there are cases with normal hemidiaphragm position on non-CT chest, when dyspnea is present clinically.

Half of the patients [47] did not recover after significant a follow-up period (1 year and more). Partial recovery was seen in significantly smaller number of patients [48], while complete recovery was present in the fewest number of patients [42].

In most cases, the type of treatment of these patients was not mentioned: Acyclovir in 2 patients, Valacyclovir in 1 patient, physical therapy in 1 patient, and topical hydrocortisone in 1 patient.

The incidence of segmental zoster abdominal paresis mimicking an abdominal hernia is relatively rare. After review from 2013 [49] with 36 patients and segmental zoster paresis of abdominal muscle (14), 8 papers with 11 patients with this problem have been published.

Their mean age (66.5 years) is not much different from the average age of the patients in the review paper from 2013 (67.5 years).

As in the review paper, the predominant level of abdominal involvement of herpes zoster was Th11.

It is the most important to exclude organic disease of the abdomen in the case of abdominal herniation when beneficial are ultrasound or CT abdomen showing normal results.

This is particularly relevant when the abdominal herniation is complicated with, for example, ileus, which is described in 1 patient.

The presence of denervation potentials in EMG of paraspinal (3 patients), or abdominal muscles (2 patients), indicates the involvement of the abdominal musculature, which separates EMG as well as a particularly useful diagnostic method in this case.

Prognosis of these patients is generally good, and there is full recovery in almost all of them in just few months [42, 43, 50].

Acyclovir is rarely used as a therapy in these patients (2 patients).

4. Discussion

Increasing rate of herpes zoster infection with increasing age particularly after age 50 years can be explained by natural decline in cell-mediated immunity to VZV with age [51]: the mean age of patients with the most frequent segmental paresis was for arms 68.56 ± 11.97 years and for legs 64.19 ± 15.28 years.

Healthy people can get HZ [31], although immunocompromised individuals are known to be at increased risk of reactivation and VZV infection [52].

The most immunocompromised patients with zoster paresis have a coexisting malignancy, diabetes mellitus, and chronic steroid therapy [53]: 7 patients suffering from lymphoma and 2 with carcinoma mammae in group with arm segmental paresis and 1 patient with segmental paresis of leg had lymphatic leukemia; as per 4 patient in each group and 1 in group with diaphragmatic paresis suffered from DM and as per 4 patient in both groups of patients with limb segmental paresis on chronic steroid therapy because of autoimmune disorders.

In clinical presentation of HZ infection, pain usually precedes the onset of the rash, and most patients have skin lesions that develop within 7 days of onset of pain [32].

Motor palsy is usually segmental, with abrupt onset reaching its maximum within a few hours [31], corresponding to the dermatomes with cutaneous lesions [18]. Interval between skin eruptions and onset of muscle weakness is generally about 2 weeks in cases with developing segmental paresis [54], but there are some reports of weakness and rash developing simultaneously [55]. With reference to the literature, maximal rash-to-weakness interval in patients with segmental limb paresis was 19 days [37] although there are different experiences. Variations in rash-to-weakness interval is best illustrated in a study of 51 patients: <3 days in 6 patients, 3–6 days in 12, 7–10 days in 14, 11–14 days in 11, 15–20 days in 5, 21–28 days in 2, and 29–35 days in 1 [33]. A delay of 4.5 months has been documented in a patient with diaphragmatic paralysis [56], when the average minimum duration of weakness was 193 days [57]. The fact that the phrenicus nerve is a motor nerve and is the longest in the body is the explanation for such a long period of time required for the development of paralysis of phrenic nerve.

Limb involvement by segmental zoster paresis is seen from 0.5 to 0.8% of all patients with cutaneous zoster [10, 11].

Upper extremity involvement is the most common region of extracranial zoster paresis involvement. There are a total of 101 patients with segmental paresis of upper limbs in recently published papers: 55 women and 45 men (**Tables 1** and **3**).

Segmental zoster paresis of the legs present in less than half the number (43 patients with leg involvement) dominated by males (20/12) (**Tables 2** and **3**).

Phrenic nerve affection was described in 26 patients (**Table 4**).

Among patients with limb paresis, proximal muscle involvement usually predominates (C5, 6, 7 or L2, 3, 4). The most commonly affected muscle is deltoid in upper limbs [78].

Besides weakness in upper limbs affected by segmental zoster paresis, in 9 patients, muscle atrophy was detected and in 3 patients in group had leg affection (in *M. quadriceps*).

It is important to recognize severe muscle weakness and atrophy in herpes zoster paresis of limb because it can be so severe to cause marked dislocation of the joint. Because of that, except pain medication, treatment for segmental paresis includes exercise that may prevent muscle atrophy and contractures [17].

Risk of postherpetic neuralgia and pain after 3 months of HZ infection increased and occurs in 8–70% of patients with HZ. People above 50 years are 15 times more likely to develop this complication [79].

In studies involving a larger number of patients with segmental zoster paresis, postherpetic neuralgia persisted in 2 and 6.6% [32, 33].

Among the patients monitored over several months and multiyear period, 13 patients with segmental zoster paresis of arm and 6 patients with paresis of leg have reported pain as the postherpetic neuralgia type (**Tables 1–3**).

The clinical diagnosis makes pain followed by rash and by weakness at the end. Once the rash appears, diagnosis of HZ can be made and laboratory confirmation is not always required.

Some patients may have zoster sine herpete, and it is zoster infection without vesicular eruptions. In making this diagnosis benefit is from varicella zoster virus antibodies in sera or cerebrospinal fluid [13]. In rare cases with herpes sine herpete, cases with prolonged period between rash and muscular weakness and cases with

Literature	Demographics, medical history	Subjective complaints	Neurological impairments	EMG findings	Other tests	Treatment	Course/outcome
[58]	40-year old women; in contact with a child with chicken pox	Rash on left buttock, burning pain in lower back, hip, and left leg	Moderate weakness in left ankle plantar flexors	Normal	Blood and CSF normal, chest and spine X-ray normal	Intrathecal Methylprednisolone, analgesics, bed rest, sodium iodide and oxytetracycline i.v.; hydrocortisone ointment for rash	Complete motor recovery after 9 months
[59]	20-year old man, chicken pox at age 7	Numbness in left foot; pain and rash over left gluteal region	Weakness in left ankle dorsiflexors	Not reported	Not reported	B1 and B12 injections	Full recovery after 2 months
[60]	31-year old man, diabetes, end stage renal failure with maternal renal allograft, autonomic neuropathy, blindness	Rash and pain in right lower back, urinary retention, weakness in right leg, diminished sensation in lumbar and sacral segments	The right ankle plantar flexors (3/5 on MRC scale) and diminished ankle jerk	Not reported	ELISA test on HZV in sera was positive for IgM and IgG, which confirmed the presence of HZ infection	Not reported	Unknown recovery of leg function, some bladder recovery
[61]	57-year old man	Rash over left T8–9 dermatomes, hyperalgesia in both legs, unable to walk, fasciculations in both legs	Flaccid paralysis of both legs, reflexes absent	Not reported	CPK: mild transient elevation; VZV complement fixation positive	Not reported	Full recovery after 3 months
[62]	70-year old woman, with hypertension	Pain in buttocks and legs; rash over left knee, thigh, buttocks; hyperesthesia in left leg; urinary incontinence	Moderate weakness in knee flexors/ extensors, mild in ankle flexors; reflexes absent in left knee, ankle	Diffuse denervation in leg and paraspinal muscles in L3–5 myotomes	X-ray: mild narrowing of L5-S1 disc space; spine MRI: right L4–5 facet joint disease	acetaminophen, oxytocin, bed rest, bupivacaine via L4–5 epidural catheter; parenteral meperidine	Near complete recovery
[63]	78-year old man, idiopathic myelofibrosis receiving cytoreductive therapy	Sudden weakness in both legs; rash in right lower leg involving knee and thigh	Bilateral leg paralysis	Not reported	CSF: increased lymphocytes, monocytes, protein, glucose; PCR on VZV	Wide spectrum antibiotic, granulocyte colony stimulating factor and erythrocyte	After initial improvement, progressive worsening

Literature	Demographics, medical history	Subjective complaints	Neurological impairments	EMG findings	Other tests	Treatment	Course/outcome
					highly positive; X-ray: ileus	transfusions; high-dose acyclovir i.v.	complicated by pneumonia and death
[64]	60-year old man with lymphatic leukemia	Burning pain along the inner aspect of the right lower leg (6/10) and rash, weakness of the right foot	Weakness of the right ankle plantar flexors (3/5 on MRC scale) and diminished ankle jerk	Fibrillations and positive sharp waves in the right gastrocnemius and paravertebral muscles (S1 root); polyphasic MUAP during activation of right foot, nerve conduction was unchanged	MRI of lumbosacral area: degenerative changes; ELISA test on VZV in sera positive for IgM and IgG	Physical therapy and oral gabapentin 900 mg/day	Motor weakness completely resolved about 6 months after the onset of neurologic symptoms but the pain was sporadic but mild
[65]	74-year old with diabetes mellitus, hypertension, and ischemic heart disease	5-day history of paraesthesia starting in the right foot and ascending up the right lower limb	Vesicular rash in the L2/3 region with MRC grading 3/5 in the right hip flexors		MRI: unremarkable	Acyclovir i.v.	Motor paresis that recovered fully with resolution of the rash
[40]	37-year old-female with history of paresis in both legs secondary to spinal cord atrophy and Vogt-Koyanagi-Harada disease and with chronic corticosteroid and azathioprine treatment of ulcerative colitis		Worsening of her baseline residual muscle strength in the right lower limb shortly after herpes zoster eruption	EMG: denervation in L3-L4 and moderate axonal polyneuropathy affecting both lower limbs			

Table 3.
Lists of studies that reviewed herpes zoster infection and motor paresis of only lower limbs.

Literature	Demographics, medical history	Subjective complaints	Neurological impairments	EMG findings	Other tests	Treatment	Course/Outcome
[66]	M, 53-year old	HZ in C3,4 dermatome	dyspnoea				
[67]	80-year old female, nephrectomy because pyelonephritis	Pain and rash over left shoulder and anterior part of chest	weakness of left sholder and proximal muscles, atrophy of supraspinatus and infraspinatus muscles; dyspnoea on left hemidiaphragm	denervation of infraspinatus and supraspinatus	Rtg- paretic left hemidiaphragm		
[44, 45]	56-year old male with peptic ulcer	rash and pain in later aspect of right arm and 3 middle fingers	general weakness of right arm and hand, atrophy of right part of neck, paralysis right hemidiaphragm		X-ray and fluoroscopy-complete paralysis of rigth hemidiaphragm		
[68]	77, F rheumatoid artritis, DM	Rash and pain in C3-5 dermatome	paralysis right hemidiaphragm				
[46]	72-year old female, RA, hypertension	Rash and pain in right C3,4 dermatome	dyspnoea				
[69]	66-year old female	HZ of left C3,4,5	after 12 months dyspnoea		X-ray –elevated left hemidiaphragm; Radioscopy-paralysis of left hemidiaphragm		
[50]	56, male bronchitis	HZ in 1st cervix dermatome	none		X-ray –elevated left hemidiaphragm		
	62-year old female astma	HZ in 1st cervix dermatome	dysponea				
	67-year old female	HZ in 1st cervix	dysponea				
	57-year old female	HZ in 1st cervix	dysponea				

Literature	Demographics, medical history	Subjective complaints	Neurological impairments	EMG findings	Other tests	Treatment	Course/Outcome
[70]	74-year old female	HZ in C3,4	dysponea				
[71]	80-year old female, hypertension, histerectomy	HZ in C3-6	Dyspnea, upper limb muscle weakness				
[72]	74-year old male, pulmonary emphysema	HZ in C2-5	deltoid muscle weakness, dyspnoea				
[16]	#2. 60-year old man with lymphatic leukemia	Rash and hypersensitivity in right C5 dermatome	coughing; enlarged liver;	Not reported	Chest X-ray: elevated right hemidiaphragm, paralysis confirmed on fluoroscopia		Complete Recovery after 1 month
[56]	74-year old male	HZ in 1st cervix	cough, dyspnea				not alleviated after 4 month
[73]	79-year old male, hypertension, carotid endarterectomy, bypass	HZ in cervix region	dyspnea, orthopnea				After 12 months not alleviated
[56]	74-year old male	HZ in cervix	cough, dyspnea				After 4 month not alleviated
[42]	A 73-year old woman	herpes zoster of left shoulder and proximal arm	weakness of left shoulder and proximal arm muscles 3 weeks after a diagnosis of herpes zoster	involvement of the C5-6 myotomes and the upper trunk of the brachial plexus	Chest X-ray and electromyographic studies documented paralysis of the left.		One year after muscle strength returned to normal, but radiographic and electrophysiologic findings of diaphragm paralysis were unchanged.

Literature	Demographics, medical history	Subjective complaints	Neurological impairments	EMG findings	Other tests	Treatment	Course/Outcome
[74]	74-year old man	HZ on left shoulder and neck	Left hemidiaphragm paralysis	Axonal changes in left nervus phrenicus	CT and X-ray- left hemidiaphragmatic relaxation	Acyclovir	not allevia after 18 months
[75]		HZ in left side of neck	Dyspnea after 3 months		X-ray- left hemidiaphragm relaxation; Pulmo-rary functions-restriction		
[76]	54-year old male	thoracic herpes zoster, 1st chest, neck, bilateral shoulders	bilateral diaphragmatic paralysis associated with brachial neuritis, orthopnea, deltoid and biceps brachii muscle weakness		Fluoroscopy		not allevia after 19 months
[42]	73-year old woman	HZ and pain of left shoulder	weakness of left shoulder and proximal arm muscles 3 weeks after HZ was diagnosed and paralysis of left hemidiaphragm	EMG – denervation, reduced recruitment od MUP Phrenic nerve conduction study- denervation without MUP	Chest X-ray- paralysis of the left diaphragm.		Weakness of arm return to normal after one year but X-ray of paralysis hemidiaphragm was unchanged
[21]	A 48-year old female	rash and vesicles over left C5-7 dermatomes	weakness in the left deltoids and biceps muscles, diminished left biceps reflex. dyspnoea with paradoxical abdominal wall movement	CT of chest normal, MRI-hyper-intensity in spinal cord at C5 level		1000 mg valacyclovir orally three times daily for seven days.	Pain resolved three months later, follow-up for 2 years without further complications.

Literature	Demographics, medical history	Subjective complaints	Neurological impairments	EMG findings	Other tests	Treatment	Course/Outcome
[77]	85-year old female breast cancer at age 84, pancreatitis, choleatitis	Rash and pain in her left neck, chest, and arm in C4,5 dermatome			A chest X-ray elevated left diaphragm	Famciclovir for 7 days at a dose of 750 mg per day.	After 14 months dyspnea and no alleviation
[48]	43-year old man	Rash on right neck and apper right hemithorax C3-5	hiccups			On X-ray paralysis of the right hemidiaphragm, HIV +	intravenous acyclovir and admitted to the hospital

Table 4.
Lists of studies that reviewed herpes zoster infection and motor paresis of diaphragmatic paralysis.

dissociation between motor segment and level of dermatomal involvement recommended laboratory confirmation of VZV infection because herpes zoster paresis may be difficult to recognize in these cases.

Possible explanation for zoster sine herpete and for herpes zoster paresis without associated dermatome eruption could be viral spread to anterior roots without corresponding axonal transport through the sensory nerve [6, 12].

Electrophysiologic study of segmental zoster paresis found reinnervation of muscles, absence of fasciculations in involved muscles, and slow motor nerve conduction velocity suggesting motor axon injury rather than anterior horn cells [80].

In electromyoneurography of 58 patients with arm paresis, denervation potential has been described in most cases (51 patients), often associated with reduced pattern and polyphasia of motor unit potentials in most cases (17 patients) and reduced motor or sensor velocity in 4 patients.

Denervation has been the most common report (in 18 patients) among patients who had electromyoneurography of leg paresis; reduced interference pattern in 12 patients and polyphasia in 4 patients were present.

Although EMNG of phrenic nerve is one of the reliable signs of affection in this nerve, it is rarely performed (1 patient).

The presence of denervation potentials in electromyography of paraspinal (3 patients), or in abdominal muscles (2 patients), indicates involvement of the abdominal musculature, which separates EMG as well as a particularly useful diagnostic method in case of abdominal wall zoster paresis (**Table 5**). Because of its self-limited nature and good prognosis, recognition of this complication is important to prevent unnecessary diagnostic studies and procedures, and because of that, abdominal hernia needs no surgery. Thus, electrodiagnostic studies can be effectively used to confirm the diagnosis.

The treatment for segmental zoster paresis includes physical therapy for weakened muscles and protection contractures with graduated exercise. This program may prevent muscle atrophy. Muscle weakness and atrophy can be so severe to cause marked dislocation of the involved joint [17].

In segmental paresis of arm, in most cases, a way of treating these patients is not mentioned. In others, physical therapy was the most common way of treating these patients (24 reported patients) and patients with leg affection, also (4 patients).

Beside physical therapy, for these patients, it is often necessary to provide pain medication, and therefore they are usually given analgesics, opioids, Amitriptyline, Pregabalin, and Gabapentin.

According to some recommendations, initiating treatment with antiviral agents as soon as the rash appears is the key to improve the outcome of herpes zoster. A 3-week course of oral corticosteroids (prednisone 60 mg/day for first week, 30 mg/day for second week, and 15 mg/day for third week) administered with the antiviral medication also has some effect on severity and duration of pain and may decrease the incidence of postherpetic neuralgia [87]. Local, epidural, and sympathetic blocks, if administered within the first 2 weeks of disease, have been reported to decrease pain and the incidence of postherpetic neuralgia [88, 89].

Initiating treatment with antiviral agents as soon as the rash appears is the key to improve the outcome of herpes zoster. As the effect of this therapy on the development of segmental paresis is not known, despite such recommendations, it appears that very small number of patients from the literature with segmental zoster paresis was treated with antiviral medication and corticosteroids. Among the antiviral drugs, Valacyclovir and Acyclovir (in total 13 patients) and corticosteroids in total 5 patients were usually used.

Previous experience in groups with a higher number of patients with segmental zoster paresis shows that the outcome of lower motor neuron involvement is

DOI: http://dx.doi.org/10.5772/intechopen.90493

Abominal wall	Demographics, medical history	Subjective complaints	Neurological impairments	EMG findings	Other tests	Treatment	Course/outcome
[81]	73-year old man with L3 vertebral compression fracture and RA (Prednisolone)	Rash and pain with blisters on his right flank	T12 and L1 segmental paresis caused abdominal wall pseudohernia, scoliosis, and standing and gait disturbance	Denervation in right T12 myotomal muscles, and MUAP markedly decreased		Orthosis Exercise	After 4 months of rehabilitation, marked improvement
[82]	72-year old man	Herpes zoster infection in T11-T12 left dermatomes	Segmental abdominal wall protrusion	Denervation in left external oblique muscle and left paraspinal muscles at T11-T12 level	SSEPs- no response in the left side at T12 dermatome		After 3 months abdominal wall protrusion had completely resolved
[83]			Abdominal wall postherpetic pseudohernia		MRI- increased signal intensity in abdominal wall muscles. Ultrasound- normal		Full recovery
[43]	62-year old male	Cutaneous vesicular eruption on the left side of the abdominal wall	Abdominal distention and paralytic ileus because of a visceral neuropathy		X-rays and CT showed distended small bowel	Acyclovir iv, oral Valacyclovir, Gabapentin 8 d after admission	Full recovery abdominal distention gradually resolved over the next 7 days and pain subsided
[47]	35 articles that described 36 individuals; mean age was 67.5 years. The ratio of men to women was 4:1.	The most affected dermatome was with rash is T11. The left and right sides were approximately equally affected	In 88.9% of the patients, herpetic rash preceded abdominal weakness. The mean latent period from rash to onset of abdominal muscle weakness was 3.5 weeks.		Electrodiagnostic studies confirmed the diagnosis in 95% of tested patients.	Conservative measures	Complete recovery with conservative measures occurred in 79.3% patients, with a mean time of 4.9 months

r-

Abominal wall	Demographics, medical history	Subjective complaints	Neurological impairments	EMG findings	Other tests	Treatment	Course/outcome
[84]	58-year old man	Rash in area 9th to 11th	Protrusion in the right abdominal wall with no pain		Ultrasonography excluded the abdominal wall defect	Oral acyclovir, mecobalamin, and vitamin B1	Disappeared after 2 months
[85]	4 patients	Involvement of posterior rami of spinal nerves in abdominal wall pseudohernia		In 3 patients, EMG of paraspinal muscles showed denervation potentials	MRI-hyperintensity of these muscles on short T1 inversion recovery imaging		
[86]	57-year-old man	Rash and dull squeezing pain	Bulge on his left flank 2 weeks after cutaneous changes				

Table 5.
Lists of studies that reviewed herpes zoster infection and motor paresis of abdominal wall.

elatively good [55]. Motor paralysis is recovered completely or nearly completely in 50–70% of cases usually within 12 months, spontaneously [11]. The time of recovery varies from 1 to 2 years [90]. Only 15% have significant deficit [33].

Among presented patients with known clinical outcomes, complete or near complete recovery is recorded in 9 patients with arm involvement and in 20 patients with leg affection.

Incomplete recovery was reported in 3 patients with arm affection and in 11 patients with leg affection (**Tables 1–3**). Two patients from each group, with segmental paresis of arms and legs, had no clinical recovery. As possible explanation of absence of a complete recovery after motor segmental paralysis caused by herpes zoster is glial scar polyradiculitis evident on MRI [91].

Prognosis in patients with diaphragmatic paralysis is not good because of prolonged reinnervation of diaphragm due to relatively long course of phrenic nerve affection, and the lack of spontaneous recovery is not surprising [74]. It is common for zoster phrenic nerve affection and associated diaphragmatic paralysis to be permanent, but occasionally, recovery has been reported after 7 and 12 months [92].

There are interesting experiences related to Piramat. It may reverse phrenic nerve paralysis in patients with diabetes, but there are no data about its use in patients with zoster-induced diaphragmatic paralysis [90].

It is important to distinguish segmental zoster paresis of abdominal wall from real abdominal wall hernia because abdominal wall hernia is mainly treated by surgery, while segmental zoster abdominal paresis needs no surgery. Ultrasonography or computed tomography (CT) is necessary to do in this case. Electromyographic testing revealed denervation in the affected dermatome and pseudohernia caused by abdominal wall paresis and is of great benefit in defining the problem and recommended in these situations.

5. Conclusion

Segmental zoster paresis is a rare complication of VZV infection. Increasing rate of herpes zoster infection and its segmental paresis is confirmed by the mean age of presented patients-it is above 65 years.

Recognizing segmental zoster paresis is important in the differential diagnosis of muscle weakness of other origin—it is of particular importance to perform electromyoneurography of paretic muscle.

Physical therapy is the most common therapy for these patients, although a lot of patients did not conducted any specific way of treatment.

Prognosis for these patients is generally good, and there is full recovery in most cases, except VZV infection of phrenic nerve and diaphragmatic paresis, where there is no significant recovery of muscle weakness in significant number of patients.

Author details

Vesna Martic
Clinic for Neurology, Military Medical Academy, Belgrade, Serbia

*Address all correspondence to: vesnamartic.bgd@gmail.com

References

[1] Peterson BH. Motor effects of herpes zoster. The Medical Journal of Australia. 1953;**2**(24):890-891

[2] Rhodes AJ, Van Rooyen CE. Textbook of Virology. Baltimore: William and Wilkins; 1962

[3] Oxman MN, Levin MJ, Johnson GR, et al. A vaccine to prevent herpes zoster and postherpetic neuralgia in older adults. The New England Journal of Medicine. 2005;**352**(22): 2271-2284

[4] Yawn BP, Saddier P, Wollan PC, St. Sauver JL, Kurland MJ, Sy LS. A population-based study of the incidence and complication rates of herpes zoster before zoster vaccine introduction. Mayo Clinic Proceedings. 2007;**82**: 1341-1349

[5] Nee PA, Lunn PG. Isolated anterior interosseus nerve palsy following herpes zoster infection: A case report and review of the literature. Journal of Hand Surgery. 1989;**14B**:447-448

[6] Hanakawa T, Hashimoto S, Kawamura J, et al. Magnetic resonance imaging in a patient with segmental zoster paresis. Neurology. 1997;**49**: 631-632

[7] Umehara T, Sengoku R, Mitsumura H, et al. Findings of segmental zoster paresis on MRI. Journal of Neurology, Neurosurgery, and Psychiatry. 2011;**82**(6):694

[8] Fabian VA, Wood B, Crowley P, Kakulas BA. Herpes zoster brachial plexus neuritis. Clinical Neuropathology. 1997;**16**(2):61-64

[9] Broadbent WH. Case of herpetic eruption in the course of branches of the brachial plexus followed by paralyisis in corresponding motor nerves. British Medical Journal. 1866;**2**:460

[10] Liu Y, Wu BY, Ma ZS, Xu JJ, Yang B, Li H, et al. A retrospective case series of segmental zoster paresis of limbs: Clinical, electrophysiological and imaging characteristics. BMC Neurology. 2018;**18**(1):121

[11] Akiyama N. Herpes zoster infection complicated by motor paralysis. Journal of Dermatology. 2000;**27**(4):252-257

[12] Cioni R, Giannini F, Passero S, et al. An electromyographic evaluation of motor complications in thoracic herpes zoster. Electromyography and Clinical Neurophysiology. 1994;**34**:125-128

[13] Bhattacharyya PK, Chakravorty NK. Lower motor neurone paralysis due to herpes zoster. The British Journal of Clinical Practice. 1988;**42**:79-82

[14] Woodland L. Herpes zoster with associated muscle paralysis. The Medical Journal of Australia. 1947;**1**(9):284

[15] Goodman CE, Kenrick MM. Herpes zoster with motor paresis. Southern Medical Journal. 1974;**67**(10):1171-1174

[16] Nord E, Weinberg A, Banjamin D, Phikhas J. Motor paralysis complicating herpes zoster. Dermatologica. 1977; **154**(5):301-304

[17] Eban R. Cervical herpes zoster and shoulder pain. British Medical Journal. 1978;**1**:177

[18] Verma AK, Maheshwari MC. Brachial monoparesis following herpes zoster. Acta Neurologica (Napoli). 1985; **7**(1):32-34

[19] Yawn BP. Post-shingles neuralgia by any definition is painful, but is it PHN? Mayo Clinic Proceedings. 2011;**86**(12): 1141-1142

[20] Yu YH, Lin Y, Sun PJ. Segmental zoster abdominal paresis mimicking an

abdominal hernia: A case report and literature review. Medicine (Baltimore). 2019;**98**(15):e15037

[21] Lin CM, Shieh WB, Chiang PC, Shieh GM, Liu Y, et al. An unusual cause of dyspnea in a patient with cervical herpes zoster. Journal of Clinical Neuroscience. 2012. DOI: 10.1016/j.jocn.2011.06.023

[22] Kang SH, Song HK, Jang Y. Zoster-associated segmental paresis in a patient with cervical spinal stenosis. The Journal of International Medical Research. 2013; **41**(3):907-913

[23] Namekawa M, Kameda T, Kumabe A, Mise J. Segmental zoster paresis of the right shoulder. Internal Medicine. 2013;**52**(24):2839

[24] Rastegar S, Mahdavi SB, Mahmoudi F, Basiri K. Herpes zoster segmental paresis in an immunocompromised breast cancer woman. Advanced Biomedical Research. 2015;**4**:170

[25] Rubin D, Fusfeld RD. Muscle paralysis in herpes zoster. California Medicine. 1965;**103**(4):261-266

[26] Pathy MS. Motor complications of herpes zoster. Age and Aging. 1979;**8**(2): 75-80

[27] Chang CM, Woo E, Yu YL, Chin D. Herpes zoster and its neurological complications. Postgraduate Medical Journal. 1987;**63**(736):85-89

[28] Rice AL, Ullal J, Vinik AI. Reversal of phrenic nerve palsy with topiramate. Journal of Diabetes and its Complications. 2007;**21**:63-67

[29] Kawajiri S, Tani M, Noda K, Fujishima K, Hattori N, Okuma Y. Segmental zoster paresis of limbs—Report of three cases and review of literature. The Neurologist. 2007;**13**(5): 313-317

[30] Greenberg J. Herpes zoster with motor involvement. JAMA. 1970; **212**(2):322

[31] Rosenfeld T, Price MA. Parilysis in herpes zoster. Australian and New Zealand Journal of Medicine. 1985;**15**: 712-716

[32] Molloy MG, Goodwill CJ. Herpes zoster and lower motor neurone paresis. Rheumatology and Rehabilitation. 1979; **18**:170-173

[33] Thomas JE, Howard FM. Segmental zoster paresis—A disease profile. Neurology. 1972;**22**(5):459-466

[34] Weiss S, Streifler M, Weiser HJ. Motor lesions in herpes zoster. European Neurology. 1975;**13**:332-338

[35] Levine M. Motor neuropathy associated with herpes zoster. Michigan Medicine. 1972;**32**:929-934

[36] Ormerod IEC, Cockrell OC. Guillian-Barre syndrome after herpes zoster infection. European Neurology. 1993;**33**:156-158

[37] Reda H, Watson J, Jones L. Zoster associated mononeuropathies: A retrospective series. Muscle & Nerve. 2012;**45**:734-739

[38] Taterka JO, Saillivan M. The motor complication of herpes zoster. Journal of the American Medical Association. 1943;**1221**:737-739

[39] Ghee LT. Herpes zoster with severe neurological complications—A report of two cases. The Medical Journal of Malaya. 1971;**26**(3)

[40] Giraldo WA, Bujidos C, Beorlegui G, San Román A, Arrebola A. Herpes zoster motor neuropathy in a patient with previous motor paresis secondary to Vogt-Koyanagi-Harada disease. American Journal of Physical

Medicine & Rehabilitation. 2013;**92**(4): 351-356

[41] Glynn C, Croxkford G, Gavaghan D, Cardo P, Price D, Miller J. Epidemiology of shingles. Journal of the Royal Society of Medicine. 1990;**83**: 617-619

[42] Bahadir C, Kalpakcioglu AB, Kurtulus D. Unilateral diaphragmatic paralysis and segmental motor paresis following herpes zoster. 2008;**38**(2): 1070-1073

[43] Anaya-Prado R, Pérez-Navarro JV, Corona Nakamura A, Anaya Fernández MM, Anaya-Fernández R, Izaguirre-Pérez ME. Intestinal pseudo-obstruction caused by herpes zoster: Case report and pathophysiology. World Journal of Clinical Cases. 2018; **6**(6):132-138

[44] Beard H. Phrenic paralysis due to herpes zoster. Case report. Medical Bulletin of US Army. 1963;**20**:106

[45] Bennett G, Watson B. Herpes zoster and postherpetic neuralgia: Past, present and future. Pain Research & Management. 2009;**14**(4):275-282

[46] Donald TC. Paralysisis of the diapragm secondary to herpes zoster. Journal of the Medical Association of the State of Alabama. 1964;**33**:306-308

[47] Chernev I, Dado DN. Segmental zoster abdominal paresis/paralysis, zoster pseudohernia or zoster lumbar hernia? Hernia. 2014;**18**(1):145-146

[48] Berger T. A rash case of hiccups. The Journal of Emergency Medicine. 2013;**44**(1):e107-e108

[49] Chernev I, Dado D. Segmental zoster abdominal paresis (zoster pseudohernia): A review of the literature. PM & R : The Journal of Injury, Function, and Rehabilitation. 2013;**5**(9):786-790

[50] Anderson JP, Keal EE. Cervical herpes zoster and dyaphragmatic paralysis. British Journal of Diseases of the Chest. 1969;**63**(4):222-226

[51] Yoshioka M, Kurita Y, Hashimoto M, Murakami M, Suzuki M. A case of segmental zoster paresis with enhanced anterior and posterior spinal roots on MRI. Journal of Neurology. 2012;**259**(3):574-575

[52] Yawn B, Gilden D. The global epidemiology of herpes zoster. Neurology. 2013;**81**(10):928-930

[53] Hilder RJ, Johnson WT. Herpes zoster with diaphragmatic paralysis and associated motor paresis. Association of Military Dermatologists. 1979;**5**:14

[54] Tilki HE, Mutluer N, Selcuki D, Stalberg E. Zoster paresis. Electromyograph. Clinical Neurophysiology. 2003;**43**:231-234

[55] Gupta SK, Helal BH, Kiely P. The prognosis in zoster paralysis. Journal of Bone and Joint Surgery. 1969;**51**(B): 593-603

[56] Stowasser M, Cameron J, Olivier WA. Diaphragmatic paralysis following cercical herpes zoster. The Medical Journal of Australia. 1990;**153**: 555-556

[57] Jones LK Jr, Reda H, Watson JC. Clinical, electrophysiologic, and imaging features of zoster-associated limb paresis. Muscle & Nerve. 2014; **50**(2):177-185

[58] Fee CF, Evarts CM. Motor paralysis of the lower extremities in herpes zoster. Cleveland Clinic Quarterly. 1968; **35**:169-176

[59] Lal S, Lebbai MS. Lower extremity paralysis complicating herpes zoster. The Journal of the Association of Physicians of India. 1970;**18**(10): 853-854

[60] Gottheiner TI, Pokroy N, Gregory MC. Herpes zoster with bladder involvement. Lancet. 1977; **1**(8010):551

[61] Chapman BA, Beaven DW. An unusual case of flaccid paralysis of both lower limbs following herpes zoster. Australian and New Zealand Journal of Medicine. 1979; **9**(6):702-704

[62] Helfgott S, Picard D, Sandeberg J. Herpes zoster radiculopathy. Spine. 1993;**18**(16):2523-2524

[63] Nicoli P, Bosa M, Rotolo A, Cilioni D, Saglio G. Herpetic leg paralysis and abdominal ileus in a patient with idiopathic myelofibrosis. Internal and Emergency Medicine. 2009;**4**:65-66

[64] Martic V. Recurrent herpes zoster with segmental paresis and postherpetic neuralgia. Vojnosanitetski Pregled. 2014;**71**(2):214-217

[65] Teo HK, Chawla M, Kaushik MA. Rare complication of herpes zoster: Segmental zoster paresis. Case Reports in Medicine. 2016:3. Article ID 7827140

[66] Halpern LS, Covner AH. Motor manifestations of herpes zoster, report of a case of associated permenent paralysis of phrenic nerve. Archives of Internal Medicine. 1949;**84**:907-916

[67] Parker G, Ramos W. Paralysis of the phrenic nerve following herpes zoster. Journal of the American Medical Association. 1962:408

[68] Spiers S. Herpes zoster and its motor lesions with a report of a case of phrenic nerve paralysis. The Medical Journal of Australia. 1963;**50**(1):850-853

[69] Brostoff J. Diaphragmatic paralysis after herpes zoster. British Medical Journal. 1966;**2**(5529):1571-1572

[70] Dutt AK. Diaphragmatic paralysis caused by herpes zoster. The American Review of Respiratory Disease. 1970; **101**(5):755-758

[71] Shivalingappa G. Diaphragmatic paralysis following herpes zoster. Gerontologia Clinica (Basel). 1970; **12**(5):283-287

[72] Derveaux L, Lacquet LM. Hemidiaphragmatic paresis after cervical herpes zoster. Thorax. 1982;**37**: 870-871

[73] Melcher WL, Die-trich RA, Whitlock WL. Herpes zoster phrenic neuritis with respiratory failure. The Western Journal of Medicine. 1990; **152**(2):192-194

[74] Soller JJ, Perpina M, Alfaro AS. Hemidiaphragmatic paralysis caused by cervical herpes zoster. Respiration. 1996;**63**:403-406

[75] Paudyal BP, Karki A, Zimmerman M, Kayastha G, Acharya P. Hemidiaphragmatic paralysis: A rare complication of cervical herpes zoster. Kathmandu University Medical Journal (KUMJ). 2006;**4**(2):246-248

[76] Hoque R, Schwendimann RN, Liendo C, Chesson AL Jr. Brachial neuritis with bilateral dia-phragmatic paralysis following herpes zoster: A case report. Journal of Clinical Neuromuscular Disease. 2008;**9**(4): 402-406

[77] Oike M, Naito T, Tsukada M, Kikuchi Y, Sakamoto N, Otsuki Y, et al. A case of diaphragmatic paralysis complicated by HZV infection. Internal Medicine. 2012;**51**:1259-1263

[78] Hope-Simpson RE. Proceedings of the Royal Society of Medicine. 1965;**58**:9

[79] Grann JW, Whytley RJ. Herpes zoster. The New England Journal of Medicine. 2002;**347**:340-346

[80] Sash GM. Segmental zoster paresis: An electrophysiological study. Muscle & Nerve. 1996;**19**(6):784-786

[81] Tashiro S, Akaboshi K, Kobayashi Y, Mori T, Naga-ta M, Liu M. Herpes zoster induced trunk muscle paresis presenting with abdominal wall pseudohernia, scoliosis, and gait disturbance and its rehabilitation: A case report. Archives of Physical Medicine and Rehabilitation. 2010;**91**(2):321-325

[82] Pérez S, Estévez NR, Conde MC. Herpes zoster-induced abdominal wall paresis: neurophysiological examination in this unusual complication. Journal of the Neurological Sciences. 2012;**312** (1–2):177-179

[83] Miranda-Merchak A, García N, Vallejo R, Varela C. MRI findings of postherpetic abdominal wall pseudo-hernia: A case report. Clinical Imaging. 2018;**50**:109-112

[84] Quin CE. Paralysis and arthropathy in her Postherpetic neuralgia (PHN), defined as pain lasting for more than 90 days after rash usually remits spontaneously, but some patients patients have pain for life

[85] Mitsutake A, Sasaki T, Hideyama T, Sato T, Katsumata J, Seki T, et al. Para-spinal muscle involvement in herpes zoster-induced abdominal wall pseudo-hernia revealed by electrophysiological and radiological studies. Journal of the Neurological Sciences. 2018;**385**:89-91

[86] Park SH, Lee SH. Segmental zoster abdominal paresis (zoster pseudohernia) preceding a skin rash. European Journal of Dermatology. 2017; **27**(5):534-535

[87] Whitley RJ, Weiss H, JW G Jr, et al. Acyclovir with and without prednisone for the treatment of herpes zoster. A randomized, placebo controlled trial. Annals of Internal Medicine (Philadelphia, PA). 1996;**125**:376-383

[88] Irving JL, Gibbons R, Meyer G, Inouye L. The effect of treting herpes zoster with oral Aciclovir in preventing postherpetic neuralgia. A meta analysis. Archives of Internal Medicine. 1997;**137**: 909-912

[89] Smith KJ, Roberts MS. Cost effectiveness of newer antiviral agents for herpes zoster: Is the evidence spotty? Infectious Diseases. 1998;**178**(Suppl): S85-S90

[90] Rice JP. Segmental zoster paresis in herpes zoster. Australian and New Zealand Journal of Medicine. 1985;**15**: 712-716

[91] Esposito MB, Arrington JA, Murtaugh FR, et al. MR of spinal cord in a patient with herpes zoster brachial plexus neuritis. Journal of Neuroradiology. 1993;**14**:203-204

[92] Oliver WA. Diaphragmatic paralysis following cervical herpes zoster. The Medical Journal of Australia. 1990;**153**: 555-556

Section 4

Human Cytomegalovirus

Chapter 5

Human Cytomegalovirus Infection: Biological Features, Transmission, Symptoms, Diagnosis, and Treatment

Şule Gökçe

Abstract

Human cytomegalovirus (CMV), a member of the human herpesviruses, is a deoxyribonucleic acid virus that is ubiquitous in the world. After primary infection, CMV develops a latent state; however, when the defense of the immune system decreases in a host, it can reactivate. Human cytomegalovirus infections are acquired via several ways. CMV is spread through contact with infected bodily fluids in humans, whereas it occurs in pregnant women through close contact with young children or through sexual transmission. The clinical manifestations consist of non-specific symptoms or clinical findings. However, the patients with acute CMV infections are generally asymptomatic. Congenital CMV infection (present at birth) occurs via intrauterine transmission of the virus that is thought to be transferred to the developing fetus. The common clinical manifestations of congenital CMV infection are sensorineural hearing loss, petechiae, jaundice at birth, and hepatosplenomegaly. The vast majority of healthy children and adolescents infected with CMV infections are most often asymptomatic. Treatment is recommended to initiate to the infants who have a symptomatic infection or primary immunodeficiency or asymptomatic infection with an isolated hearing loss. The diagnosis of congenital CMV infections should be considered when it is detected in the newborns with signs and symptoms consistent with congenital CMV disease or with abnormal neuroimaging consistent with CMV or newborns who have documented sensorineural hearing loss.

Keywords: human cytomegalovirus, transmission, symptoms, diagnosis

1. Introduction

Human cytomegalovirus (CMV), an infectious agent that is ubiquitous in the world population, is a member of human herpesvirus family including viruses such as Epstein-Barr virus, herpes simplex virus, and varicella zoster. CMV is the largest virus of the herpes family with a size of 190 nm. The structure of the virus includes a complex envelope composed of host cell-derived membrane studded and an icosahedral capsid that contains the virion DNA. The tegument layer which describes an amorphous area between the envelope and the capsid induces strong adaptive immune responses including CMV-specific CD8+ cytotoxic T lymphocytes that are believed to play a fundamental role in controlling CMV replication in the infected host. Humans are the only source of cytomegalovirus, and it might infect all races, ages, and genders.

Human cytomegalovirus, an opportunistic pathogen, is transmitted through solid organ/bone transplantation, placental pathway, sexual intercourse, blood transfusion, and in close contact with virus-spreading people. Similar to other herpesviruses, it leads to primary infection, secondary infection, latent infection, and reactivation or reinfection. In immunocompromised patients, CMV is responsible for serious clinical symptoms. After primary infection, the viral genome forms an episomal circular form in myeloid progenitors and endothelial cells. Via TNF-alpha and type II interferon (IFN), latent CMV alter to active CMV and migrate to inflamed tissue where they further propagate during an inflammatory process and a context of immune activation. CMV is a viral agent that also causes intrauterine infections and brings about deafness and neurological anomalies in newborns. Cytomegalovirus has been associated with variable seroprevalence from 45 to 100% in women of reproductive age and is never cleared from the host [1]. In immunocompetent adults, primary CMV infections are usually asymptomatic. Less commonly, fever, lymphadenopathy, and mononucleosis-like syndrome coexist with peripheral lymphocytosis. When people with impaired immune function are not treated, it can lead to infections with high mortality [2].

2. Epidemiology

Human cytomegalovirus infections are acquired via several ways. CMV is generally transmitted by infected fluids (e.g., saliva, breast milk, blood products) contact at home and nursery schools as community exposure. It is also thought that the contaminated urine contact might be a role of cytomegalovirus transmission [3]. Among the postnatal contaminations, breastfeeding is the most common course of CMV infection in young infants particularly in populations with high CMV seroprevalence and high rates of breastfeeding from seropositive women during the first 6 months of breastfeeding, peaking at 4–8 weeks after delivery, but the risk continues for the duration of breastfeeding. Even though it is thought that a local reactivation in the mammary glands of the mothers with latent CMV infection can cause the transmission to baby, but the transmission mechanism has not been clearly defined yet [4]. As a reservoir of virus, these infants excrete virus in the saliva and urine for prolonged periods to other infants, children, and adults. The preschool children with CMV can disseminate through stool, and they might infect their parents and teachers in baby care centers. Throughout childhood and early adulthood, CMV is transmitted by exposure to saliva and urine. Due to the fact that the virus is present in seminal and cervical fluids, it can also be transmitted by sexual way.

Congenital CMV infection (present at birth) occurs via intrauterine transmission of the virus that is thought to be transferred to the developing fetus in approximately 30% of women undergoing primary infection during pregnancy or by reactivation in women previously immune for CMV rate on the order of 1–2% (vertical road). In the USA, Canada, Western Europe, and Australia, it is a common infection that is estimated to occur in about 5–7 per 1000 live births [5]. Rates as the highest with 2% in Asia and Africa have been described. Congenital CMV infection contributes to permanent disabilities such as hearing loss, vision loss, cerebral palsy, and/or cognitive impairment in children. Approximately 90% of newborn with congenital CMV are asymptomatic; however, the newborns are also at risk for CMV-associated disabilities [6].

Nosocomial infections with CMV emerge from exposure to blood products containing CMV. Transfusion-acquired CMV infections often caused symptomatic illness including hepatitis and thrombocytopenia on children and adults. Fatal infections might be developed in newborns who are being born from women without immunity of CMV due to lacking antibodies of CMV. Severe infection can also present in immunocompromised patients who received blood with CMV. Recently, using blood products from CMV-seronegative donors, the incidence of transfusion-associated CMV infections has greatly decreased. Infections arising from CMV transferred in the allograft are major causes of morbidity in the early and late period after transplantation. Even if the antiviral therapy is applied, CMV infection causes long-term graft dysfunction and graft loss, particularly in cardiac and lung transplant recipients [7].

3. Clinical manifestations

The clinical manifestations consist of non-specific symptoms or clinical findings. However, the patients with acute CMV infections are generally asymptomatic. During the intrauterine period, babies infected with CMV usually do not expose symptoms at birth. However, 10–15% of them exhibit symptoms and become symptomatic at a later stage of their infancy. The common clinical manifestations of congenital CMV infection are sensorineural hearing loss, petechiae, jaundice at birth, hepatosplenomegaly, small size for gestational age, microcephaly, lethargy and/or hypotonia, poor suck, chorioretinitis, seizures, hemolytic anemia, and pneumonia [8]. The sensorineural hearing loss may be detectable at birth; however, approximately 30% of the cases have delayed onset. **Table 1** shows clinical manifestations, treatment, and outcome of congenital CMV infections. Congenital CMV associated with hearing loss is bilateral in 71% of children. Eye examination of symptomatic congenital CMV includes chorioretinitis that is the most common ocular abnormality, retinal scars, optic atrophy, central vision loss, and strabismus. The endocrinologic manifestations consist of Graves' disease and diabetes insipidus and renal disease, such as nephrotic syndrome. Congenital CMV infections are the leading cause of other long-term neurodevelopmental disabilities and can also affect other organ systems such as gastrointestinal system including Menetrier disease. Ascites, myocarditis, cardiomyopathy, ventricular trabeculations, and enterocolitis are also seen among manifestations of the congenital CMV in symptomatic neonates [9]. Despite the fact that the overall mortality rate among infants with congenital CMV infection is approximately 4–8% within the first year of life, congenital CMV infection might be associated with mortality in premature infants and infants with primary immune disorders of T cells or natural killer cells. Rarely, many infants without underlying disease

	Clinical manifestations	Treatment	Outcome
Congenital CMV infections	**At birth, 90% of cases are asymptomatic** Petechiae, jaundice at birth, hepatosplenomegaly, petechial rash, small size for gestational age, thrombocytopenia, microcephaly, intracranial calcifications, polymicrogyria, ventriculomegaly, sensorineural hearing loss, chorioretinitis, seizures	Asymptomatic infants do not require antiviral treatment Ganciclovir or valganciclovir for symptomatic infections	Overall mortality rate is 4–8% Mortality rate with severe fulminant disease is as high as 30%. Long-term sequelae include hearing loss, cerebral palsy, intellectual disability, vision impairment, and seizures

Table 1.
Clinical manifestations, treatment, and outcome of congenital CMV infections.

have the risk of mortality as high as 30% caused by a fulminant course. Death is usually caused by viral-associated hemophagocytic syndrome or severe end-organ disease of the liver, lungs, bone marrow, or central nervous system. Even if the infants with a life-threatening disease associated with congenital CMV are able to live, neurological sequelae might persist for a lifetime (e.g., microcephaly, intellectual disability, cerebral palsy, and hearing disorders) [10, 11]. Pneumonitis, signs of viral sepsis, thrombocytopenia, and coinfections are more likely to have existed in premature infants than term neonates [7]. Additionally, preterm infants can also present with a classical triad of apneas, bradycardia, and gray pallor describing CMV sepsis-like syndrome [12]. The cases with perinatal infections which are generally not associated with any clinical manifestations can be acquired during birth or ingestion of CMV-containing breast milk. Disseminated infections associated with end-organ disease and death might be seen in extremely premature infants or infants born to nonimmune women.

The vast majority of healthy children and adolescents infected with CMV infections (acquired CMV infection) are most often asymptomatic (**Tables 2** and **3**). However, the remaining patients (approximately 10%) could present with several mild or moderate symptoms. A clinical entity named mononucleosis-like syndrome with no heterophile antibody titers or positive monospot tests is characterized by fever, fatigue, pharyngitis, adenopathy (especially cervical adenopathy), and hepatitis. A headache, abdominal pain with diarrhea, arthralgias, and rash can also be observed in mononucleosis-like syndrome in acquired CMV infection. Laboratory findings mimic to EBV-related mononucleosis syndrome like lymphocytosis or lymphopenia with thrombocytopenia and elevated transaminases [13]. Unusual manifestations or complications of acquired CMV infections in healthy individuals consist of pneumonitis, myopericarditis, hemolytic anemia, viral hemophagocytic syndrome, granulomatous hepatitis, Guillain-Barré syndrome, and meningoencephalitis [7].

When a reactivation of the endogenous virus and infection from the transplanted organ or from blood product transfusion take place in immunocompromised children and adolescents, serious CMV disease can be observed which is linked to the underlying disease process responsible for the immunosuppression (**Table 4**). Non-specific symptoms such as fever, malaise, and leukopenia might be seen in all types of patients. Patients with renal transplant are at graft loss; the liver recipients are most likely to be associated with hepatitis and colitis. Early myocarditis followed by late atherosclerosis is seen in heart transplant recipients. Recipients

	Clinical manifestations	Treatment	Outcome
Early postnatal CMV infection	**Term infants: Most infants are asymptomatic** Fever, hepatosplenomegaly, mild pneumonitis, abnormal blood counts, abnormal liver function tests **Premature and VLBW infants: Infection can be severe and life-threatening** Sepsis-like syndrome, hepatosplenomegaly, pneumonitis, hepatitis, NEC, abnormal blood counts	Most term infants and asymptomatic preterm infants do not require antiviral treatment Ganciclovir or valganciclovir for severe symptomatic infections in premature infants	Term infants: no permanent sequelae. Premature and VLBW infants: mortality rate with symptomatic infection is 5–10% There does not appear to be increased risk of hearing loss, cerebral palsy, or other neurodevelopmental disabilities; however, long-term outcomes are not clearly understood

Table 2.
Clinical manifestations, treatment, and outcome of early-postnatal CMV infection.

	Clinical manifestations	Treatment	Outcome
CMV infection in immunocompetent children and adolescents	**Most children are asymptomatic** Fever, fatigue, pharyngitis, mononucleosis-like syndrome, adenopathy, hepatitis, headache, abdominal pain, diarrhea, arthralgias, rash, lymphocytosis or lymphopenia, thrombocytopenia, abnormal liver function tests, negative monospot test	Antiviral treatment is generally not indicated Supportive care with hydration and fever control	No permanent sequelae

Table 3.
Clinical manifestations, treatment, and outcome of CMV infection in immunocompetent children and adolescents.

	Clinical manifestations	Treatment	Outcome
CMV infection in immunocompromised children and adolescents	**Infection can be severe and life-threatening** Fever, malaise, leukopenia, hepatitis, pneumonitis, colitis, graft loss (in patients with organ transplant), myocarditis, retinitis, encephalitis/encephalopathy (especially in patients with HIV)	Ganciclovir or valganciclovir	High risk of morbidity and mortality; depends in part on the underlying condition

Table 4.
Clinical manifestations, treatment, and outcome of CMV infection in immunocompromised children and adolescents.

with lung and bone marrow transplant generally show a state of pneumonia. Children with human immunodeficiency virus (HIV) and CMV coinfection tend to have retinitis, colitis, pneumonitis, and encephalitis/encephalopathy [14].

4. Diagnosis

The diagnosis of congenital CMV infections should be considered when it is detected in the newborns with signs and symptoms consistent with congenital CMV disease or with abnormal neuroimaging consistent with CMV or newborns who have documented sensorineural hearing loss (SNHL). Newborns who are possessed by a mother with seroconversion positivity or with positive CMV immunoglobulin G (IgG) and CMV immunoglobulin M (IgM) antibody or with mononucleosis-like illness during pregnancy should also be evaluated for congenital CMV infections. Infants with an abnormal T-cell receptor in newborn screening should be suspected for congenital CMV infections. The reliable diagnosis of congenital CMV infections must be detected through sources of virus and viral nucleic acids from urine, saliva, and blood within the first 3 weeks of life in infants. The newborns with suspected congenital CMV infection should be analyzed by viral culture, modified culture (also called rapid culture or shell vial assay), and polymerase chain reaction (PCR) for diagnosis. Detection of CMV by PCR in blood or plasma samples is more accurate than other diagnostic tests [15]. Due to the fact that the polymerase chain reaction provides quantitative results, urine and saliva that include high levels of CMV DNA are generally evaluated in newborns suspected with CMV.

Primary infection of CMV in nonimmunocompromised individuals requires evidence of an IgM reactivity for CMV in blood. It can persist for months depending on the sensitivity of the particular assay. Due to the fact that the infected individuals can intermittently shed virus, body fluids such as saliva or urine do not provide the diagnosis of CMV infection. Polymerase chain reaction-based methods are also utilized for diagnosis of urine, saliva, and blood and in tissue specimens obtained at biopsy. A combination with immunofluorescence detection of CMV-encoded and conventional culture of CMV using human dermal fibroblasts remains standard in many institutions. Characteristic nuclear (and cytoplasmic) inclusions (owl's eye inclusions) can be detected by histologic stains in tissue specimens.

5. Treatment

Treatment is recommended to be initiated to the infants who have a symptomatic infection or primary immunodeficiency or asymptomatic infection with an isolated hearing loss for congenital CMV infection within the first month of life. At the first line, ganciclovir through intravenous route and its orally available prodrug, valganciclovir, for treatment of congenital CMV disease are indicated. Randomized clinical trials of the Collaborative Antiviral Study Group suggested that 6 weeks of ganciclovir treatment could confine hearing loss and improve developmental outcome in infants infected symptomatically. In addition, infants with severe perinatal CMV infection caused by breast milk could be treated with ganciclovir [7, 16].

In the normal host and early-postnatal infections in term asymptomatic infants, treatment of acquired CMV infections is not recommended. However, treatment must be implemented in preterm or very-low-birth-weight (VLBW) infants with severe infection.

Immunocompromised hosts are recommended with antiviral therapy with ganciclovir in children with active CMV disease. Even if there are no signs or symptoms of end-organ disease in immunocompromised children, it is also suggested to apply the antiviral therapy due to CMV viremia (e.g., positive or rising quantitative PCR). Treatment of CMV disease in immunocompromised pediatric patients consists of ganciclovir and oral valganciclovir similar to that in adults. Treatment is usually given in the form of intravenous ganciclovir or oral valganciclovir for 4–6 weeks on average [17]. Patients are recommended to seek for antiviral treatment monitorization with white blood cell count, AST, and ALT regularly. In some cases, when the antiviral therapy cannot improve clinical or virologic response within several weeks of treatment with ganciclovir or valganciclovir, foscarnet might be integrated to the treatment. Another specific anti-CMV treatment is cidofovir that may be used in children with careful monitoring of renal function and metabolic condition when antiviral resistance occurs.

6. Prevention

6.1 Passive immunoprophylaxis

Intrauterine disease could be prevented from infection through passive transfer of anti-CMV antibodies. However, this transfer is not successful in allograft recipients. A neonatal transfer is prevented from CMV infection using CMV-seronegative or leukocyte-reduced blood for extremely preterm infants. Freezing/thawing or pasteurization eliminates the risk of transfusion-related CMV infection in breast milk contaminated with CMV, however; it might not be complete. The passive transfer of anti-CMV antibodies is recommended to limit transmission and disease in pregnant women with primary CMV infection.

6.2 Active immunoprophylaxis

A number of different vaccine platforms that will be available for prevention of CMV disease in newborns and immunocompromised patients have been evaluated in clinical trials. However, none are licensed.

7. Final considerations

Cytomegalovirus is a common agent that can be seen all over the world. Although CMV is generally asymptomatic in healthy individuals, it can lead to severe infection in individuals with immune deficiency. During pregnancy, CMV also cause significant disabilities in babies whose mother has CMV. After its primary infection, the virus remains as an inactive form in the host throughout the life. Recurrent CMV infections may occur with the reactivation of the silent virus. Due to the fact that congenital CMV infections are asymptomatic, a hearing test is strongly recommended to the newborns.

Author details

Şule Gökçe
Ege University Faculty of Medicine, Department of Pediatrics, General Pediatrics Unit, Ege University, Izmir, Turkey

*Address all correspondence to: sule.gokce@ege.edu.tr

References

[1] Cannon MJ, Schmid DS, Hyde TB. Review of cytomegalovirus seroprevalence and demographic characteristics associated with infection. Reviews in Medical Virology. 2010;**20**:202-213. DOI: 10.1002/rmv.655

[2] Crumpacker CS, Wadhwa S. Cytomegalovirus. In: Mandell GL, Bennett JE, Dolin R, editors. Mandell, Douglas and Bennett's Principles and Practice of Infectious Diseases. 6th ed. Philadelphia: Churchill Livingstone; 2005. pp. 1786-1801

[3] Lanzieri TM, Dollard SC, Josephson CD, et al. Breast milk-acquired cytomegalovirus infection and disease in VLBW and premature infants. Pediatrics. 2013;**131**:1937-1945. DOI: 10.1542/peds.2013-0076

[4] Hamprecht K, Maschmann J, Jahn G, et al. Cytomegalovirus transmission to preterm infants during lactation. Journal of Clinical Virology. 2008;**41**:198-205. DOI: 10.1016/j.jcv.2007.12.005

[5] Kenneson A, Cannon MJ. Review and meta-analysis of the epidemiology of congenital cytomegalovirus (CMV) infection. Reviews in Medical Virology. 2007;**17**:253-276. DOI: 10.1002/rmv.535

[6] Britt WJ. Cytomegalovirus. In: Remington J, Klein J, Wilson C, Nizet V, Maldonado Y, editors. Infectious Diseases of the Fetus and Newborn Infant. 7th ed. Philadelphia, PA: Elsevier Saunders; 2011. pp. 706-755

[7] Harrison GJ. Cytomegalovirus. In: Cherry JD, Harrison GJ, Kaplan SL, et al, editors. Feigin and Cherry's Textbook of Pediatric Infectious Diseases. 7th ed. Philadelphia: Elsevier Saunders; 2014. p. 1969

[8] Dreher AM, Arora N, Fowler KB, et al. Spectrum of disease and outcome in children with symptomatic congenital cytomegalovirus infection. The Journal of Pediatrics. 2014;**164**:855-859. DOI: 10.1016/j.jpeds.2013.12.007

[9] Jin HD, Demmler-Harrison GJ, Coats DK, et al. Long-term visual and ocular sequelae in patients with congenital cytomegalovirus infection. The Pediatric Infectious Disease Journal. 2017;**36**:877-882. DOI: 10.1097/INF.0000000000001599

[10] Britt W. Cytomegalovirus. In: Wilson CB, Nizet V, Maldonado YA, Remington JS, Klein JO, editors. Remington and Klein's Infectious Diseases of the Fetus and Newborn Infant. 8th ed. Philadelphia: Elsevier Saunders; 2016. p. 724

[11] Snider M, Noyola D, Griesser C, Demmler GJ. Congenital cytomegalovirus disease (C-CMV-D) registry 1990-2007: Targets for treatment and prevention revealed. In: Abstract Presentation at Pediatric Academic Societies Annual Meeting; 1 May 2008; Honolulu, Hawaii. 2008

[12] Kurath S, Halwachs-Baumann G, Muller W, Resch B. Transmission of cytomegalovirus via breast milk to the prematurely born infant: A systematic review. Clinical Microbiology and Infection. 2010;**16**:1172-1178. DOI: 10.1111/j.1469-0691.2010.03140.x

[13] American Academy of Pediatrics. Cytomegalovirus infection. In: Kimberlin DW, editor. Red Book: 2015 Report of the Committee on Infectious Diseases. 30th ed. Elk Grove Village, IL: American Academy of Pediatrics; 2015. p. 317

[14] Kovacs A, Schluchter M, Easley K, et al. Cytomegalovirus infection and HIV-1 disease progression in infants born to HIV-1-infected women. Pediatric pulmonary and cardiovascular

complications of vertically transmitted HIV infection study group. The New England Journal of Medicine. 1999;**341**:77-84. DOI: 10.1056/NEJM199907083410203

[15] Pinninti SG, Ross SA, Shimamura M, et al. Comparison of saliva PCR assay versus rapid culture for detection of congenital cytomegalovirus infection. The Pediatric Infectious Disease Journal. 2015;**34**:536-537. DOI: 10.1097/INF.0000000000000609

[16] Kimberlin DW, Lin CY, Sánchez PJ, et al. Effect of ganciclovir therapy on hearing in symptomatic congenital cytomegalovirus disease involving the central nervous system: A randomized, controlled trial. The Journal of Pediatrics. 2003;**143**:16-25. DOI: 10.1016/S0022-3476(03)00192-6

[17] Asberg A, Rollag H, Hartmann A. Valganciclovir for the prevention and treatment of CMV in solid organ transplant recipients. Expert Opinion on Pharmacotherapy. 2010;**11**:1159-1166. DOI: 10.1517/14656561003742954

Section 5

Epstein-Barr Virus

Chapter 6

The Role of the Epstein-Barr Virus Lytic Cycle in Tumor Progression: Consequences in Diagnosis and Therapy

Emmanuel Drouet

Abstract

The Epstein-Barr virus (EBV) reactivation corresponds to the activation of EBV global replication involving not only the origin of the latent viral replication but also that of the origin of lytic replication. During this reactivation, a minority of B cells infected with EBV in its latent form enter the lytic phase. During this phase, all EBV proteins are produced, enabling the assembly of complete virions that lysate their host cells and infect neighboring cells (lytic cycle). This horizontal EBV transmission seeks to increase the pool of EBV-infected B cells. This chapter seeks to review the role of the lytic EBV proteins (particularly that of the ZEBRA protein) *in tumor development*. This protein is the main transcription factor of EBV, expressed during the activation of the lytic cycle. Recently, we demonstrated that this immediate early protein can be detected in the soluble state (s-ZEBRA) in the serum of patients with posttransplant lymphoproliferative disorder. We highlighted the role of ZEBRA in EBV pathogenesis in transplanted subjects, not only as a key protein in the activation of EBV replication but also as a protein "toxoid" released into the extracellular milieu. This release could result in increased secretion of immunomodulatory cytokines and that of angiogenesis-promoting factors conducive to tumor progression.

Keywords: tumor progression, tumorigenesis, EBV, lytic cycle, Zta/ZEBRA protein, lymphoma

1. Introduction

Epstein-Barr virus (EBV) is a member of the herpesvirus family and only infects primates, with tropism for B cells and epithelial cells, which establishes a lifelong persistent infection in over 90% of the world's population [1]. After the resolution of the primary infection episode, EBV enters a latent phase. Following primary infection, the virus establishes lifelong persistence within the host memory B-cell compartment utilizing restricted latent gene expression programs [2–4]. Like all herpesviruses, EBV can choose between two alternative lifestyles: latent or lytic replication. EBV lytic replication, which is required for horizontal spread of the virus from cell to cell, and from host to host, occurs in both epithelial cells and B cells [5–7]. During latency, EBV exists in a dormant state where only a viral gene subset is expressed, facilitating the episomal persistence of the viral genome [8]. However, under certain physiological conditions, both viruses undergo lytic reactivation leading to expression of the full lytic gene complement, followed by the assembly and egress of infectious virions [1].

Lytic replication is mediated by the virally encoded DNA polymerase using the oriLyt replication origin and results in the release of infectious viral particles [9]. In contrast, during latent viral infection, only a viral gene subset is expressed. The genome is replicated once per cell cycle using the cellular DNA polymerase and the oriP replication origin, and the progeny virus is not released. Latent EBV infection allows the virus to persist for the life of the host and to avoid detection by the immune system [1]. In immune-competent hosts, the outgrowth of EBV-immortalized B cells is prevented through a robust expansion of EBV-specific memory T cells directed toward both lytic and latent antigens [10]. Thus, EBV-positive individuals with either congenital or acquired immune deficiencies are highly susceptible to viral reactivation and malignant transformation. Furthermore, EBV-seropositive patients treated with immunosuppressive medications following organ/stem cell transplantation are at significant risk of developing posttransplant lymphoproliferative disorder (PTLD), an often fatal B-cell lymphoproliferative disease (LPD) [11, 12].

Importantly however, EBV (as other gammaherpesviruses, e.g., KSHV) can also undergo abortive lytic reactivation, resulting in the expression of early lytic genes without subsequent virion assembly and cell lysis [13]. Although numerous efforts to understand these disorders' molecular basis have focused on viral latency, EBV lytic cycles are now widely accepted as major contributors to oncogenesis and could be important targets in the development of new therapeutic or diagnostic approaches [14, 15]. Thus, in this chapter, we will discuss how lytic replication, particularly via the immediate-early/early antigens or IEAs/EAs, augments the pathogenesis of EBV-associated malignancies, especially in the lymphoma setting or with respect to the treatments that potentially target the lytic replication cycle.

2. Arguments of the pathogenic role of lytic infection in EBV-associated malignancies

EBV establishes a lifelong infection in B lymphocytes achieved through a highly regulated viral gene expression program. Numerous studies have been focused on the contributions of EBV latent infection in the pathogenesis of EBV-induced malignancies. These studies have revealed that the membrane LMP1 protein is the major essential oncoprotein [1, 16]. In recent years, the viral lytic cycle was shown to play an essential role in carcinogenesis through several potential mechanisms. EBV lytic infection may increase the total number of latently infected cells by enhancing transmission of the virus from cell to cell and thus constitutes an essential aspect of viral pathogenesis. A small subset of lytically infected cells was detected in biopsies of EBV-associated malignancies [13, 17–22], suggesting a potential role of viral lytic infection in promoting tumor growth in vivo. Furthermore, several studies have indicated that the viral lytic cycle in a fraction of B cells promotes the transformation of B lymphocytes in vitro [23] and growth of B-cell lymphoma in vivo [24, 25] through the release of paracrine growth factors and angiogenic factors [26]. Focusing on the study of NPC, Wu et al. revealed that recurrent EBV reactivation promotes genome instability, invasiveness, and tumorigenesis of NPC cells. Another Wu et al. finding was that the contribution of the lytic cycle is more marked than that of the latent infection [27, 28]. Additionally, lytic replication enhances in NPC cells the secretion of vascular endothelial growth factor (VEGF), a factor contributing to angiogenesis and subsequent nasopharyngeal carcinoma's

(NPC) metastasis or relapse after remission [29]. Clinical and epidemiological studies have revealed that individuals with elevated plasma EBV DNA load and antibody titers against the lytic viral capsid antigen (VCA) and early antigen (EA) display a high risk of NPC [30]. These studies have also shown that fluctuation of EBV antibody titers occurs prior to the onset of NPC [31–34]. Attention has also focused on the diagnostic and prognostic value of IgG antibody against the IEA ZEBRA (see further) in patients with NPC [33–37] and in Hodgkin disease (HD) patients as well [38]. More recently we reported that the replicative form of EBV, as investigated using anti-ZEBRA IgG titers, correlates with poorer outcomes in women with breast cancer [39]. All in all, these studies investigating serological lytic EBV markers have emphasized the prominent role importance of the lytic infection in EBV malignancies. Recently, the identification by mass spectrometry of 44 different EBV proteins in Burkitt lymphoma cells undergoing EBV replication has increased the knowledge base on EBV lytic replication, possibly highlighting different targets for future therapeutic strategies [40].

3. The role of some lytic EBV proteins in the tumorigenesis and focusing onto the Zta/ZEBRA IE transactivator

The role of the lytic EBV proteins was highlighted in some recent reviews [13, 14]. Overall, the authors clearly demonstrated the role of these lytic EBV proteins in tumorigenesis. They also showed that EBV reactivation may aid virus transmission within the tumor microenvironment to establish latency and drive cellular proliferation. Having said that, the likely predominant role of the EBV lytic cycle is likely to provide the necessary paracrine, anti-apoptotic, and immunomodulatory signals required for tumorigenesis (**Table 1**). From this point on, we will focus on the critical role of the IE ZEBRA *trans*-activator, because this protein has a key role in the balance between the EBV latent and lytic cycles. The switch from latent to lytic infection is mediated by the IE protein ZEBRA (Orf BZLF1) and R (Orf BRLF1) [6, 41–45]. ZEBRA and R proteins are transcription factors that activate one another's promoters, as well as their own promoters [8]. In combination, ZEBRA and R induce expression of all early (E) lytic viral proteins, thereby enabling the viral replication. The ZEBRA protein belongs to the bZIP family of transcription factors and is homologous to c-jun and c-fos which bind to the consensus AP1 motif as well as to atypical AP1-like motifs known as Z-responsive elements (ZREs) [42, 44, 46]. The EBV genome is not methylated in virions. However, in cells with long-term latent infection, the majority of the EBV genome proves to be highly methylated [47–51]. DNA methylation, which plays a critical role in modulating the expression of both cellular and viral genes, induces transcriptional repression through multiple different mechanisms including prevention of transcription factor binding to DNA [48, 52–56]. Surprisingly, while DNA methylation of the EBV IE BRLF1 promoter (Rp) inhibits its activation through cellular transcription factors, it enhances ZEBRA's ability to activate the R [48]. This unusual effect of Rp methylation on ZEBRA activation is due to the enhanced ability of ZEBRA to bind to the methylated, versus unmethylated, forms of two atypical CpG-containing Rp ZRE sites and requires serine residue 186 in the basic DNA domain of ZEBRA [48, 49]. Several properties of ZEBRA are listed in **Table 2**. ZEBRA was shown to inhibit antiviral cytokine signaling [57, 58] and disrupt T-cell recognition of MHC-II molecules [59]. As ZEBRA is able to activate host cellular genes (i.e., the immunomodulatory genes IL-10 and IL-13 [60, 61]), it was also found that EBV reactivation may contribute to the growth of latently infected cells [25, 62] by promoting the release of B-cell growth factors [26].

EBV lytic gene	Cellular homologs	Lytic function	Mechanisms of oncogenesis	Oncogenic function	Clinical and diagnostic relevance (Ref)
ZEBRA (Zta, BZLF1)	c-Fos, c-Jun	IEA, transactivator in cooperation with Rta	Promotes secretion of proangiogenic factors, VEGF and cytokines (IL-6, IL-8, IL-10, IL-13), toxoid protein	Angiogenesis, B-cell growth, immunomodulation	Expression in PTLD, B-cell lymphomas, Hodgkin lymphomas, NPC, detection of s-ZEBRA in PTLD patients' serum, specific antibodies in NPC/HD/PTLD patients [84]
BHRF1 and BALF1	Bcl-2	EA, delay cell death	Inhibition of apoptosis	Cell survival	
BILF1	G protein-coupled receptor	EA immunoevasin	MHC class I downregulation		
BNRF1		EA nucleocapsid trafficking	Interacts with centromeres to cause centrosome overduplication	Genome instability	
BALF3	Terminase	EA DNA synthesis and incorporation into virions	DNA damage	Genome instability	
BGLF4		EA viral DNA replication and nuclear import	DNA damage	Genome instability	
BARF1	C-fms receptor	EA immunomodulator	Modulates expression genes involved in apoptotic signaling	Cell survival	Most abundant EBV transcripts in NPC, detection of s-BARF1 in the NPC patients' serum [110] and specific antibodies [111]
BGLF5		Host cell shutoff (DNase), TLR-9 downregulation	DNA damage	Genome instability	Expression in NPC biopsies [112]. Specific antibodies in NPC [113] patients
BcRF1	Il-10	Immunoevasin	Downregulates interferon gamma	Cell survival	Soluble vIL10 in NHL patients' serum [114]

EA, early antigen; IEA, immediate-early antigen; HD, Hodgkin disease; NHL, non-Hodgkin lymphoma; NPC, nasopharyngeal carcinoma; PTLD, posttransplant lymphoproliferative disorder.

Table 1.
EBV lytic proteins and tumorigenic functions (summarized from Ref. [14]).

Functions of ZEBRA protein	Target	Effect
ZEBRA mediates the switch from latent to lytic infection, in combination with the second transactivator Rta	Virus	ZEBRA and Rta induce expression of all early (E) lytic viral proteins, allowing the virus to replicate [45]
ZEBRA is able to bind to the methylated DNA	Virus	Activation of EBV replication [48]
ZEBRA exhibits cell-penetrating properties, like HIV-1 Tat and dengue virus NS1	Host cell	Toxoid effect [82–84]
Activation of immunomodulatory genes IL-10 and IL-13	Host cell	ZEBRA contributes to the growth of latently infected cells by promoting the release B-cell growth factors [60, 61]
Inhibition of antiviral cytokine signaling	Immune system	Immune subversion [57]
High immunogenicity: ZEBRA elicits robust B- and T-cell responses that dominate the early immune responses in patients	Immune system	EBV homeostasis and immune control of EBV replication activation [67–69] Detection of specific IgG antibodies in EBV-associated diseases [34, 38, 75, 115, 116]
ZEBRA promotes B-cell transformation and lymphomagenesis in a humanized NOD/LtSz-scid/IL2Rγnull mouse model of EBV-related disease (PTLD)	Animal model	ZEBRA might be a prominent factor in the development of EBV-associated malignancy [24, 62]
ZEBRA expression (mainly through abortive cycles) was shown in EBV-associated tumors, including Hodgkin disease and PTLD	Host tissue	Pathogenesis and tumor progression in patients [13, 17, 94, 96]
ZEBRA is also released in the extracellular milieu (as HIV-1 Tat) and detectable in the serum of transplant patients, particularly those with PTLD	Host tissue	Diagnostic and prognostic value [84]

Table 2.
Characteristics of the ZEBRA transactivator.

Scid/hu mice proved to be a good model to study the pathogenesis of posttransplant B-cell lymphomas [63–65]. Several recent studies emphasized the role of EBV lytic proteins (including ZEBRA) in promoting B-cell transformation and lymphomagenesis in a humanized NOD/LtSz-scid/IL2Rγ^{null} mouse model of an EBV-related disease [24]: The authors modeled a humanized NOD/LtSz-scid/IL2Rγ^{null} mouse that was reconstituted with CD34+ human hematopoietic progenitor cells isolated from fetal liver and also simultaneously xenotransplanted with fetal liver and thymus tissues [hNSG(thy)]. The hNSG(thy) mice were infected with EBV developing CD20-positive diffuse large B-cell lymphomas (DLBCL). To understand the contribution of the lytic cycle in lymphomagenesis, the hNSG(thy) mice were infected with a BZLF1 knockout EBV. Although no significant differences were apparent in the number of infected cells and in their tissue distribution, lymphomas were found in 6/11 (54.4%) of the mice infected with the wild-type virus and in only 2/14 (14.3%) of mice infected with the mutant virus. These data along with other studies support that the lytic cycle directly helps tumorigenic mechanisms rather than the lateral spread of viruses, thereby increasing the number of latently infected cells [66]. In line with this observation, lymphomas developing in the mice infected with wild-type viruses sometimes expressed ZEBRA and BMRF1 lytic proteins but never gp350/220 structural protein. Thus, these data support the participation of the abortive lytic cycle in lymphomagenesis (see below). Hong et al. made a similar observation [62]: This research group reported that when lymphoblastoid cell lines (LCLs) derived from wild type and virus knockout of BZLF1 and BRLF1 IE genes were transplanted into scid mice, both mutant LCLs were shown unable to induce lymphoproliferative disease, supporting a critical role of the lytic cycle in the EBV-induced lymphoproliferation [62].

It appears clearly that ZEBRA expression is critical for viral activation, persistence, and disease pathogenesis. From an immunologic standpoint, ZEBRA expression is thus likely to represent a prime target of adaptive immune responses (T-cell and humoral responses) (**Table 2**). ZEBRA, highly immunogenic, elicits robust T-cell responses [10, 67–72] that dominate the early immune responses in patients [70]. B-cell epitopes were additionally described; to illustrate, the DNA-binding domain of ZEBRA (basic region, including the so-called RAK epitope) is a major target antigen for IgM antibody response in EBV primary infection (45), whereas the N-terminus part (activation domain) is mainly recognized by IgG in patients with EBV reactivation [35, 73–75].

4. ZEBRA, a deleterious protein acting as a "toxoid"

ZEBRA appears to be a multifunctional protein controlling its own expression, virus replication, cell cycle arrest, and DNA damage response in the host cells [76, 77–80]. The structure of the 245-amino acid ZEBRA protein has been resolved in its DNA-bound form [81]. Based on this finding, the protein appears divided into an N-terminal *trans*-activation region and a basic DNA-binding domain flanked by a coiled-coil dimerization region (zipper) [81]. The C-terminal domain interacts with the zipper region by forming intra- and intermolecular interactions, resulting in a hydrophobic pocket. These complex interactions are unique among the bZIP members and result in the stabilization of the ZEBRA dimer when bound to DNA [81] (**Figure 1**).

In 2005, we demonstrated that this multifunctional protein not only binds DNA and controls virus replication but also has the ability to penetrate lymphoid cells [82]. Moreover, we also demonstrated that ZEBRA possesses a transduction domain (C-terminus), crosses the outer membrane of live cells, and accumulates in the nucleus of lymphocytes. The protein transduction mostly occurs via direct translocation through the lipid bilayer rather than by endocytosis [83]. Soluble ZEBRA has additionally been detected in the sera of posttransplant lymphoproliferative disease patients, suggesting a possible secretion of this protein into the circulating

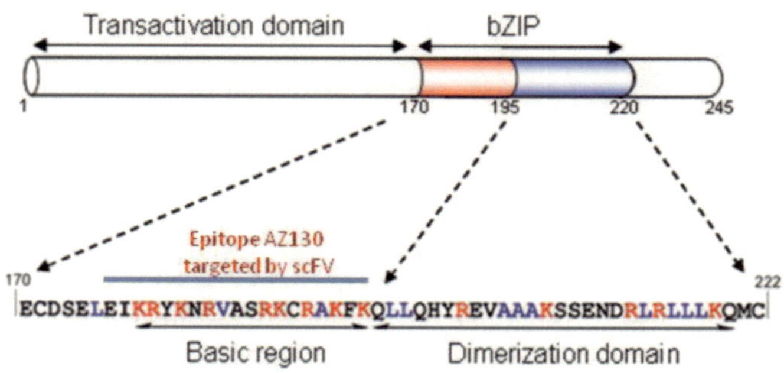

Figure 1.
Schematic representation of the ZEBRA protein. The cell-penetrating peptide is located at the C-terminus (aa 178–220), enabling the cell transduction (see Ref. [83]). Two mAbs AZ125 and AZ130 specific for ZEBRA protein (recognizing the transactivation domain and the DNA-binding domain, respectively) were employed in a sandwich ELISA throughout the whole procedure for quantifying captured s-ZEBRA [84]. The DNA-binding domain of ZEBRA (basic region, including the so-called RAK epitope) is targeted by AZ130 mAb-derived scFv and could be used as possible neutralizing therapeutic drug.

blood, with further evidence toward a "toxoid" role for this lytic protein [84]. This unusual property must be compared with those of other nonstructural viral proteins like HIV-1 Tat and dengue virus (DV) NS1 [85, 86]. Therefore, we hypothesized that ZEBRA is to be involved in tumorigenesis and tumor progression, not only as an essential protein for EBV replication activation but also as a "toxin" released in the extracellular milieu. We hypothesized that early abortive replication associated with fully lytic cycles may occur within the tumor or its environment, along with eventual ZEBRA release in the bloodstream (**Figure 2**). In the end, this phenomenon could lead to the secretion of cytokines and angiogenesis-promoting factors, further aggravating the immunosuppressive environment (**Figure 3**) [26, 87].

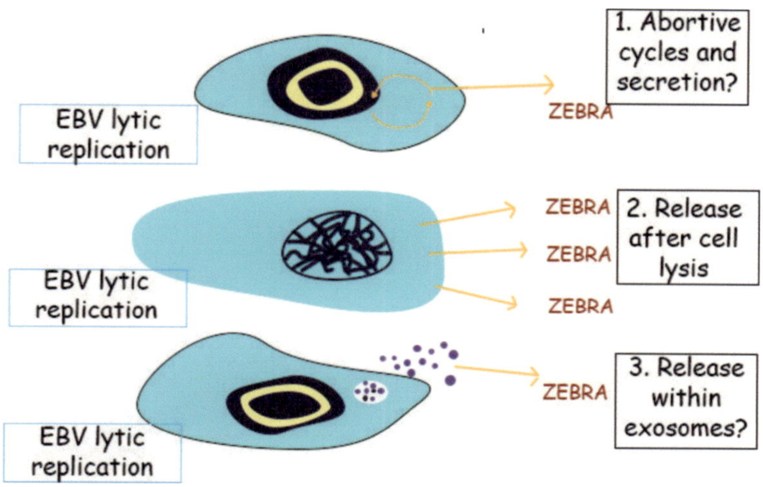

Figure 2.
Schematic representation of different possibilities of ZEBRA's release from the EBV-infected cells.

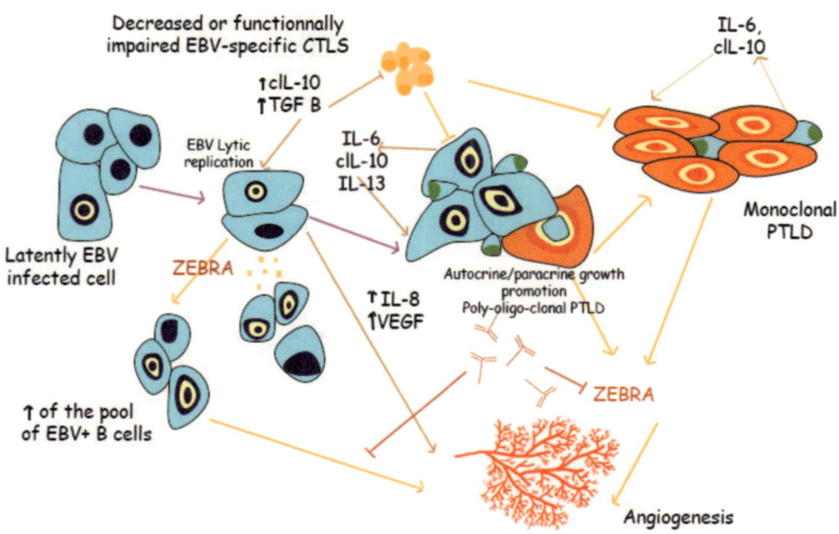

Figure 3.
Multiple effects of the ZEBRA protein on tumorigenesis and tumor progression. At any level, ZEBRA is likely to be expressed and released in the circulation as s-ZEBRA. The expression of ZEBRA and EBV replication activation may be modulated by T-cell and B-cell responses in the context of lytic/abortive cycles. There is possibility to neutralize s-ZEBRA by IgG antibodies, and ZEBRA protein could be a novel EBV-specific antigen suitable for immunotherapeutic approach.

5. Evidence of abortive lytic cycle and their role in tumorigenesis

In the absence of other lytic genes, particularly those encoding late structural proteins, without the formation of infective viral particles, BZLF1 expression is termed the "abortive lytic cycle" (**Figure 4**) [13]. The existence of abortive cycles was demonstrated in EBV-associated malignancies through the detection of either the ZEBRA protein (via monoclonal antibodies) or mRNA: Hodgkin disease [17], Non-Hodgkin lymphoma (NHL) [18, 88], NPC [20], or Burkitt lymphoma [21]. Decades ago, we revealed the early stages of EBV replication in lymphomas in scid/hu mice, assessed by the expression of ZEBRA expression, whereas the VCA expression late replicate protein proved to be weak [89]. In a recent review, the authors discussed evidence supporting an abortive lytic cycle with several lytic genes expressed, such as immunomodulatory (BCRF1, BARF1, BNLF2A, BGLF5, and BILF1) and anti-apoptotic (BHRF1 and BALF1) proteins. In their paper, the authors also discussed how the EBV immunomodulatory mechanisms result in paracrine signals that feed tumor cells. The existence of such abortive lytic cycles would reconcile the requirement of lytic expression in view of viral tumorigenesis without relying on a complete cycle that would induce cell lysis, thus releasing the newly formed infective viral particles [13].

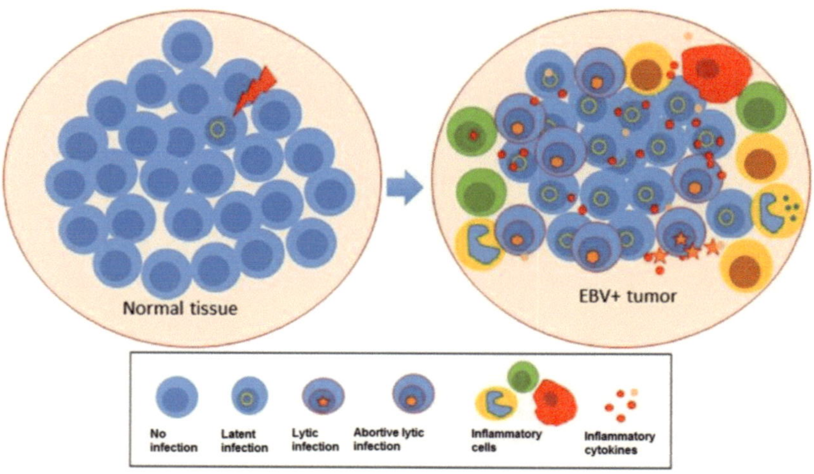

Figure 4.
The existence of abortive lytic cycles (presence of ZEBRA protein, with absence of other lytic genes, particularly those encoding late structural proteins) would reconcile the requirement of lytic expression in view of viral tumorigenesis without relying on a complete lytic (from Ref. [13] with permission).

6. Consequences in terms of diagnosis and therapy

The observation of ZEBRA transduction properties and abortive cycles has prompted us to imagine the existence of a circulating ZEBRA form, especially in the serum of patients, particularly those with lymphoid tumors. Many articles have previously mentioned the existence of the lytic cycle, especially in transplant patients with PTLD. Studies conducted on EBV lytic proteins, especially the IE proteins like ZEBRA, in patients with PTLD or HIV-associated NHL are still scarce, mostly relating to the role of EBV proteins and gene products in neoplastic tissues [18, 40, 90–95]. It must be pointed out that several authors exploring BZLF1 transcripts in the peripheral blood lymphocytes (PBL) of PTLD patients demonstrated that both a high EBV genome number and strong BZLF1 mRNA expression are sensitive markers of EBV-related PTLD [96]. In a previous study, we demonstrated ZEBRA expression in the whole peripheral blood mononuclear cells (PBMCs) from a patient exhibiting a LPD using flow cytometry. In these patients who underwent non-myeloablative allogeneic stem cell transplantation, the ZEBRA antigen was found in mostly 5% of PBMCs [97]. Moreover infected cells were detected in the peripheral blood at higher levels (e.g., 1–10 lytic-infected cells per 10^4 B lymphocytes [1, 98] versus 1 and 50 per 10^6 B cells in persistently infected healthy individuals) [99].

Recently we succeeded in detecting soluble ZEBRA (s-ZEBRA) protein in serum from transplant patients (measured by an antibody-based ELISA). The s-ZEBRA (>100 ng/mL) was predictive in 80% of PTLD patients within 10 weeks, prior to PTLD diagnosis ($p < 0.0001$) [84]. We applied this technique in both solid organ transplant patients and in hematopoietic stem cell (HSC) patients. During the HSCT patient follow-up, the availability of iterative serum samples enabled us to investigate the kinetics of s-ZEBRA appearance in comparison to that of EBV DNA qPCR and anti-ZEBRA IgG antibodies. As for patient follow-up, it was interesting to notice that the circulating ZEBRA protein could be detected during periods in which the viral DNA was not detectable by qPCR. This could be explained by certain inconsistencies observed between the qPCR and s-ZEBRA detection results (**Figure 5**). This discrepancy may be accounted for by the precocity of the ZEBRA signal measured over the course of EBV infection in this patient population. This precocity of s-ZEBRA detection (with respect to the qPCR) was independent of the PCR format, since we observed the same phenomenon in PTLD patients who were investigated by measuring the EBV load (expressed in copies/150,000 cells). In two PTLD patients, s-ZEBRA was detected at 2 and 6 weeks, respectively, prior to the PTLD episode and before the increase in qPCR signals [100]. It is interesting to note that the s-ZEBRA potentially correlated the symptomatology, as only one patient (#P3) (**Figure 5**) exhibited very high levels (3690 ng/mL) compared to the two others (#P4 and #P10) without any PTLD (162 and 300 ng/mL, respectively) (**Table 3**).

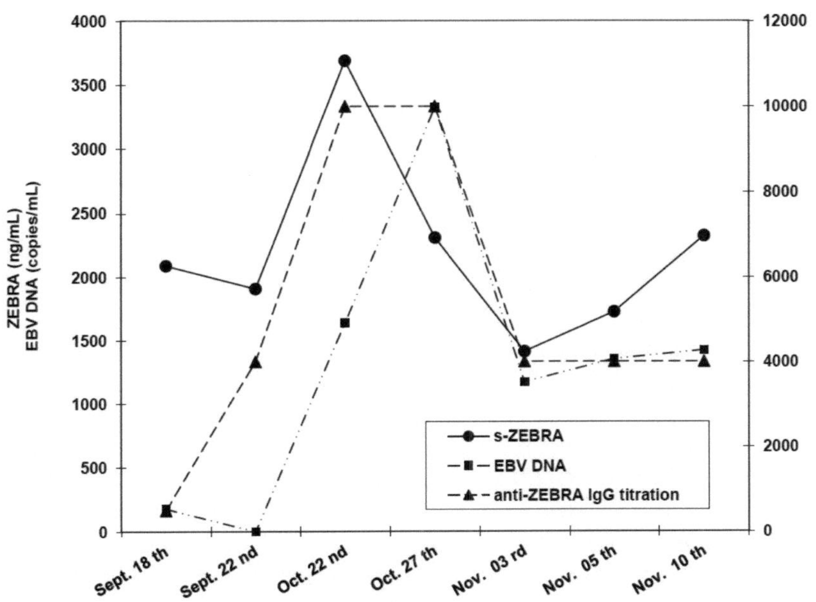

Figure 5.
*Follow-up of the HSCT patients #P3 (see **Table 3**): Levels of s-ZEBRA (ng/mL), titers of anti-ZEBRA IgG, and EBV DNA load (copies/mL) are noted, respectively. This patient experienced a lymphoproliferative episode (Oct. 27) and then was treated by anti-CD20 therapy, with decrease of the markers explored. The re-increase in s-ZEBRA during the course of the GvHD is worth noticing, until the patient finally died (Nov. 10).*

Patients	Nr of serum samples	Means (m)			PTLD
		s-ZEBRA (OD_{450} value)	anti-ZEBRA (IgG titer)	EBV DNA load (copies/mL)	
Immunocompetent Seronegative	30	0.067	NT	0	No
Immunocompetent Seropositive	25	0.092	NT	NT	No
Infectious mononucleosis	50	0.238	24,000	NT	No
Hematopoietic stem cell transplant patients					
P1	3	0.461	2833	93	No
P2	3	0.321	3333	28	
P3 (deceased)	7	3.26	5214	1300	Yes
P4	3	0.92	2000	947	No
P5	4	0.274	2300	136	No
P6	3	0.255	200	2417	No
P7	2	0.24	2250	42	No
P8	2	0.201	4000	117	No
P9	2	0.248	4000	NT	No
P10	1	1.236	20,000	0	No
P11	1	0.323	20,000	0	No
	31	m = 0.727	m = 6012	m = 507.9	

Only one patient (#P3) experienced a PTLD episode. For patients #P3, #P4, and #P10, the OD_{450} values correspond to 3690, 162, and 300 ng/mL of circulating s-ZEBRA, respectively. NT, not tested. EBV DNA load in serum was measured by qPCR (see Ref. [84]).

Table 3.
Summary of the s-ZEBRA ELISA, anti-ZEBRA IgG titration, and EBV PCR assay results in the control populations (immunocompetent EBV-seronegative/EBV-seropositive subjects, patients with primary infection) and in 11 hematopoietic stem cell transplant patients.

Therefore, s-ZEBRA detection could be a potential diagnostic marker for EBV follow-up in immunocompromised patients. Given this particular setting, our findings suggest that s-ZEBRA testing could help identify patients likely to develop severe outcomes during the critical posttransplant period. Based on our findings, we assume that the circulating ZEBRA form to be a useful target for the rapid and early diagnosis of PTLD, meaning that ZEBRA antigen-capture ELISA is likely to be a good confirmatory test for assessing EBV load in this patient population. Future evaluations of the prognostic value of ZEBRA detection should focus on the sampling time and clinical phase of lymphoproliferative disease. To improve the sensitivity and specificity of PTLD diagnosis [101–103], we hypothesize that combining approaches to detect both the circulating antigen and EBV load would be likely to increase the sensitivity and reliability of tests designed to identify such malignant EBV-related diseases.

These data focused on the relevance of the lytic cycle have already attracted the attention of the EBV community due to the potential usefulness of targeting certain lytic proteins (**Figure 6**). Investigations using both in vitro and in vivo systems revealed that FDA-approved leflunomide, a teriflunomide metabolite that targets EBV replication, inhibited the earliest step of lytic EBV reactivation (BZLF1 and BMRF1 expression) and thus prevented the development of EBV-induced lymphomas in both a humanized mouse model and a xenograft model [104].

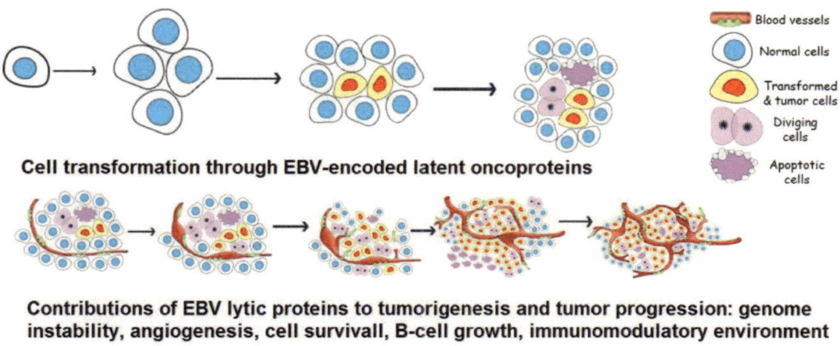

Figure 6.
Contributions of latent and lytic EBV in tumorigenesis and tumor progression. Consequences in clinical settings: In PTLD patients, EBV DNA quantification (qPCR) in blood samples corresponds to the burden of memory B cells (not proliferating lymphoblasts) infected by EBV. Thus, the increased EBV load in PBMCs, currently measured in the routine patients' follow-up, is accounted for by an increase in the number circulating EBV-positive cells (memory B cells), similar to latently infected resting B cells. Conversely, the measurement of circulating s-ZEBRA could be a novel biomarker for PTLD prognosis. The potential usefulness of targeting certain lytic proteins is also critical for managing several EBV malignancies in the future.

More recently duvesilib (a molecule inhibiting the PIU3K/akt signaling pathway, thereby inhibiting BCR signaling) was shown to reduce the expression of EBV lytic genes like BZLF1 and gp350/220, in EBV-positive cell lines and cell growth, suggesting that this molecule was able to suppress the lytic EBV cycle induced by BCR signaling [105]. The histone acetylase and DNA methyl transferase inhibitors are possible avenues to suppress the ZEBRA expression and entire lytic cascade [106]. Immunotherapeutic approaches such as vaccination against IE proteins or IE-specific therapeutic monoclonal antibodies (mAbs) look likewise promising. A recent study demonstrated that vaccination of hu-PBL-SCID mice against the ZEBRA protein could enhance specific cellular immunity and significantly delay the development of lethal EBV-LPD [107]. Efforts are additionally being made to improve the quality of CD4+ T-cell line infusions responding to EBV lytic antigens [108]. Recently authors demonstrated the role of BARF1 as a novel EBV-specific antigen suitable for immunotherapeutic approach. These authors provided evidence that mABs anti-BARF1 are likely to be a potent tool for managing several EBV malignancies [109].

In conclusion, the relevance of the lytic cycle and, particularly, the role of ZEBRA in lymphomagenesis is a new paradigm pertaining to the prevention and treatment strategies for EBV-associated cancers. Therefore, it now appears relevant to investigate the lytic EBV infection in immunocompromised patients, such as organ transplant recipients, who are highly prone to developing EBV-associated malignancies. With respect to circulating s-ZEBRA, we have made the following assumptions: (i) it may be a marker of over-immunosuppression by triggering the expression of immunomodulating cytokines; (ii) it may thus consequently play a specific role in the oncogenic process, even tumor progression. More efforts should be invested to examine the potential of drugs that target EBV lytic proteins, especially the IE proteins, such as ZEBRA.

Author details

Emmanuel Drouet
HIV and Persisting Viruses Lab, Institut de Biologie Structurale, Grenoble-Alpes University, Grenoble, France

*Address all correspondence to: emmanuel.drouet@ibs.fr

References

[1] Longnecker RM, Kieff E, Cohen JI. Epstein-Barr virus. In: Knipe DM, Howley PM, editors. Fields Virology. Philadelphia, Pa: Lippincott-Williams Wilkins Publishers; 2013. pp. 1898-1960

[2] Joseph AM, Babcock GJ, Thorley-Lawson DA. EBV persistence involves strict selection of latently infected B cells. Journal of Immunology. 2000;**165**:2975-2981

[3] Babcock GJ, Decker LL, Volk M, Thorley-Lawson DA. EBV persistence in memory B cells in vivo. Immunity. 1998;**9**:395-404

[4] Rowe M, Lear AL, Croom-Carter D, Davies AH, Rickinson AB. Three pathways of Epstein-Barr virus gene activation from EBNA1-positive latency in B lymphocytes. Journal of Virology. 1992;**66**:122-131

[5] Li QX, Young LS, Niedobitek G, Dawson CW, Birkenbach M, Wang F, et al. Epstein-Barr virus infection and replication in a human epithelial cell system. Nature. 1992;**356**:347-350

[6] Takada K, Shimizu N, Sakuma S, Ono Y. Trans activation of the latent Epstein-Barr virus (EBV) genome after transfection of the EBV DNA fragment. Journal of Virology. 1986;**57**:1016-1022

[7] Laichalk LL, Thorley-Lawson DA. Terminal differentiation into plasma cells initiates the replicative cycle of Epstein-Barr virus in vivo. Journal of Virology. 2005;**79**:1296-1307

[8] Kenney SC, Mertz JE. Regulation of the latent-lytic switch in Epstein-Barr virus. Seminars in Cancer Biology. 2014;**26**:60-68

[9] Tsurumi T, Fujita M, Kudoh A. Latent and lytic Epstein-Barr virus replication strategies. Reviews in Medical Virology. 2005;**15**:3-15

[10] Rickinson AB, Long HM, Palendira U, Munz C, Hislop AD. Cellular immune controls over Epstein-Barr virus infection: New lessons from the clinic and the laboratory. Trends in Immunology. 2014;**35**:159-169

[11] Cesarman E. Gammaherpesvirus and lymphoproliferative disorders in immunocompromised patients. Cancer Letters. 2011;**305**:163-174

[12] Dierickx D, Habermann TM. Post-transplantation lymphoproliferative disorders in adults. The New England Journal of Medicine. 2018;**378**:549-562

[13] Morales-Sanchez A, Fuentes-Panana EM. The immunomodulatory capacity of an Epstein-Barr virus abortive lytic cycle: Potential contribution to viral tumorigenesis. Cancers (Basel). 2018;**10**:1-25

[14] Manners O, Murphy JC, Coleman A, Hughes DJ, Whitehouse A. Contribution of the KSHV and EBV lytic cycles to tumourigenesis. Current Opinion in Virology. 2018;**32**:60-70

[15] Jha HC, Banerjee S, Robertson ES. The role of gammaherpesviruses in Cancer pathogenesis. Pathogens. 2016;**5**:1-43

[16] Young LS, Rickinson AB. Epstein-Barr virus: 40 years on. Nature Reviews. Cancer. 2004;**4**:757-768

[17] Brousset P, Knecht H, Rubin B, Drouet E, Chittal S, Meggetto F, et al. Demonstration of Epstein-Barr virus replication in Reed-Sternberg cells of Hodgkin's disease. Blood. 1993;**82**:872-876

[18] Brousset P, Drouet E, Schlaifer D, Icart J, Payen C, Meggetto F, et al. Epstein-Barr virus (EBV) replicative gene expression in tumour cells of AIDS-related non-Hodgkin's lymphoma in relation to CD4 cell number and antibody titres to EBV. AIDS. 1994;**8**:583-590

[19] Bibeau F, Brousset P, Knecht H, Meggetto F, Drouet E, Rubin B, et al. Epstein-Barr virus replication in Hodgkin disease. Bulletin du Cancer. 1994;**81**:114-118

[20] Martel-Renoir D, Grunewald V, Touitou R, Schwaab G, Joab I. Qualitative analysis of the expression of Epstein-Barr virus lytic genes in nasopharyngeal carcinoma biopsies. The Journal of General Virology. 1995;**76**(Pt 6):1401-1408

[21] Xue SA, Lu QL, Poulsom R, Karran L, Jones MD, Griffin BE. Expression of two related viral early genes in Epstein-Barr virus-associated tumors. Journal of Virology. 2000;**74**:2793-2803

[22] Cabras G, Decaussin G, Zeng Y, Djennaoui D, Melouli H, Broully P, et al. Epstein-Barr virus encoded BALF1 gene is transcribed in Burkitt's lymphoma cell lines and in nasopharyngeal carcinoma's biopsies. Journal of Clinical Virology. 2005;**34**:26-34

[23] Katsumura KR, Maruo S, Takada K. EBV lytic infection enhances transformation of B-lymphocytes infected with EBV in the presence of T-lymphocytes. Journal of Medical Virology. 2012;**84**:504-510

[24] Ma SD, Hegde S, Young KH, Sullivan R, Rajesh D, Zhou Y, et al. A new model of Epstein-Barr virus infection reveals an important role for early lytic viral protein expression in the development of lymphomas. Journal of Virology. 2011;**85**:165-177

[25] Ma SD, Yu X, Mertz JE, Gumperz JE, Reinheim E, Zhou Y, et al. An Epstein-Barr virus (EBV) mutant with enhanced BZLF1 expression causes lymphomas with abortive lytic EBV infection in a humanized mouse model. Journal of Virology. 2012;**86**:7976-7987

[26] Dolcetti R, Dal Col J, Martorelli D, Carbone A, Klein E. Interplay among viral antigens, cellular pathways and tumor microenvironment in the pathogenesis of EBV-driven lymphomas. Seminars in Cancer Biology. 2013;**23**:441-456

[27] Fang CY, Lee CH, Wu CC, Chang YT, Yu SL, Chou SP, et al. Recurrent chemical reactivations of EBV promotes genome instability and enhances tumor progression of nasopharyngeal carcinoma cells. International Journal of Cancer. 2009;**124**:2016-2025

[28] Wu CC, Liu MT, Chang YT, Fang CY, Chou SP, Liao HW, et al. Epstein-Barr virus DNase (BGLF5) induces genomic instability in human epithelial cells. Nucleic Acids Research. 2010;**38**:1932-1949

[29] Hong GK, Kumar P, Wang L, Damania B, Gulley ML, Delecluse HJ, et al. Epstein-Barr virus lytic infection is required for efficient production of the angiogenesis factor vascular endothelial growth factor in lymphoblastoid cell lines. Journal of Virology. 2005;**79**:13984-13992

[30] Chien YC, Chen JY, Liu MY, Yang HI, Hsu MM, Chen CJ, et al. Serologic markers of Epstein-Barr virus infection and nasopharyngeal carcinoma in Taiwanese men. The New England Journal of Medicine. 2001;**345**:1877-1882

[31] Ji MF, Wang DK, Yu YL, Guo YQ, Liang JS, Cheng WM, et al. Sustained elevation of Epstein-Barr virus antibody levels preceding clinical onset of nasopharyngeal carcinoma. British Journal of Cancer. 2007;**96**:623-630

[32] Cao SM, Liu Z, Jia WH, Huang QH, Liu Q, Guo X, et al. Fluctuations of Epstein-Barr virus serological antibodies and risk for nasopharyngeal carcinoma: A prospective screening study with a 20-year follow-up. PLoS One. 2011;**6**:e19100

[33] Dardari R, Hinderer W, Lang D, Benider A, El Gueddari B, Joab I, et al. Antibody responses to recombinant Epstein-Barr virus antigens in nasopharyngeal carcinoma patients: Complementary test of ZEBRA protein and early antigens p54 and p138. Journal of Clinical Microbiology. 2001;**39**:3164-3170

[34] Dardari R, Khyatti M, Benider A, Jouhadi H, Kahlain A, Cochet C, et al. Antibodies to the Epstein-Barr virus transactivator protein (ZEBRA) as a valuable biomarker in young patients with nasopharyngeal carcinoma. International Journal of Cancer. 2000;**86**:71-75

[35] Dardari R, Menezes J, Drouet E, Joab I, Benider A, Bakkali H, et al. Analyses of the prognostic significance of the Epstein-Barr virus transactivator ZEBRA protein and diagnostic value of its two synthetic peptides in nasopharyngeal carcinoma. Journal of Clinical Virology. 2008;**41**:96-103

[36] Yip TT, Ngan RK, Lau WH, Poon YF, Joab I, Cochet C, et al. A possible prognostic role of immunoglobulin-G antibody against recombinant Epstein-Barr virus BZLF-1 transactivator protein ZEBRA in patients with nasopharyngeal carcinoma. Cancer. 1994;**74**:2414-2424

[37] Mathew A, Cheng HM, Sam CK, Joab I, Prasad U, Cochet C. A high incidence of serum IgG antibodies to the Epstein-Barr virus replication activator protein in nasopharyngeal carcinoma. Cancer Immunology, Immunotherapy. 1994;**38**:68-70

[38] Drouet E, Brousset P, Fares F, Icart J, Verniol C, Meggetto F, et al. High Epstein-Barr virus serum load and elevated titers of anti-ZEBRA antibodies in patients with EBV-harboring tumor cells of Hodgkin's disease. Journal of Medical Virology. 1999;**57**:383-389

[39] Marrao G, Habib M, Paiva A, Bicout D, Fallecker C, Franco S, et al. Epstein-Barr virus infection and clinical outcome in breast cancer patients correlate with immune cell TNF-alpha/IFN-gamma response. BMC Cancer. 2014;**14**:665

[40] Traylen C, Ramasubramanyan S, Zuo J, Rowe M, Almohammad R, Heesom K, et al. Identification of Epstein-Barr virus replication proteins in Burkitt's lymphoma cells. Pathogens. 2015;**4**:739-751

[41] Rooney CM, Rowe DT, Ragot T, Farrell PJ. The spliced BZLF1 gene of Epstein-Barr virus (EBV) transactivates an early EBV promoter and induces the virus productive cycle. Journal of Virology. 1989;**63**:3109-3116

[42] Farrell PJ, Rowe DT, Rooney CM, Kouzarides T. Epstein-Barr virus BZLF1 trans-activator specifically binds to a consensus AP-1 site and is related to c-fos. The EMBO Journal. 1989;**8**:127-132

[43] Speck SH, Chatila T, Flemington E. Reactivation of Epstein-Barr virus: Regulation and function of the BZLF1 gene. Trends in Microbiology. 1997;**5**:399-405

[44] Flemington E, Speck SH. Evidence for coiled-coil dimer formation by an Epstein-Barr virus transactivator that lacks a heptad repeat of leucine residues. Proceedings of the National Academy of Sciences of the United States of America. 1990;**87**:9459-9463

[45] Feederle R, Kost M, Baumann M, Janz A, Drouet E, Hammerschmidt W, et al. The Epstein-Barr virus lytic program is controlled by the co-operative functions of two transactivators. The EMBO Journal. 2000;**19**:3080-3089

[46] Flemington E, Speck SH. Autoregulation of Epstein-Barr virus putative lytic switch gene BZLF1. Journal of Virology. 1990;**64**:1227-1232

[47] Lee S, Cho HJ, Park JJ, Kim YS, Hwang S, Sun R, et al. The ORF49 protein of murine gammaherpesvirus 68 cooperates with RTA in regulating virus replication. Journal of Virology. 2007;**81**:9870-9877

[48] Bhende PM, Seaman WT, Delecluse HJ, Kenney SC. The EBV lytic switch protein, Z, preferentially binds to and activates the methylated viral genome. Nature Genetics. 2004;**36**:1099-1104

[49] Bhende PM, Seaman WT, Delecluse HJ, Kenney SC. BZLF1 activation of the methylated form of the BRLF1 immediate-early promoter is regulated by BZLF1 residue 186. Journal of Virology. 2005;**79**:7338-7348

[50] Ernberg I, Falk K, Minarovits J, Busson P, Tursz T, Masucci MG, et al. The role of methylation in the phenotype-dependent modulation of Epstein-Barr nuclear antigen 2 and latent membrane protein genes in cells latently infected with Epstein-Barr virus. The Journal of General Virology. 1989;**70**(Pt 11):2989-3002

[51] Minarovits J, Minarovits-Kormuta S, Ehlin-Henriksson B, Falk K, Klein G, Ernberg I. Host cell phenotype-dependent methylation patterns of Epstein-Barr virus DNA. The Journal of General Virology. 1991;**72**(Pt 7): 1591-1599

[52] Bird AP, Wolffe AP. Methylation-induced repression–Belts, braces, and chromatin. Cell. 1999;**99**:451-454

[53] Boyes J, Bird A. Repression of genes by DNA methylation depends on CpG density and promoter strength: Evidence for involvement of a methyl-CpG binding protein. The EMBO Journal. 1992;**11**:327-333

[54] Buschhausen G, Wittig B, GraessmannM, GraessmannA. Chromatin structure is required to block transcription of the methylated herpes simplex virus thymidine kinase gene. Proceedings of the National Academy of Sciences of the United States of America. 1987;**84**:1177-1181

[55] Clark SJ, Harrison J, Molloy PL. Sp1 binding is inhibited by (m)Cp(m)CpG methylation. Gene. 1997;**195**:67-71

[56] Doerfler W. DNA methylation and gene activity. Annual Review of Biochemistry. 1983;**52**:93-124

[57] Morrison TE, Mauser A, Wong A, Ting JP, Kenney SC. Inhibition of IFN-gamma signaling by an Epstein-Barr virus immediate-early protein. Immunity. 2001;**15**:787-799

[58] Morrison TE, Kenney SC. BZLF1, an Epstein-Barr virus immediate-early protein, induces p65 nuclear translocation while inhibiting p65 transcriptional function. Virology. 2004;**328**:219-232

[59] Zuo J, Thomas WA, Haigh TA, Fitzsimmons L, Long HM, Hislop AD, et al. Epstein-Barr virus evades CD4+ T cell responses in lytic cycle through BZLF1-mediated downregulation of CD74 and the cooperation of vBcl-2. PLoS Pathogens. 2011;7:e1002455

[60] Mahot S, Sergeant A, Drouet E, Gruffat H. A novel function for the Epstein-Barr virus transcription factor EB1/Zta: Induction of transcription of the hIL-10 gene. The Journal of General Virology. 2003;**84**:965-974

[61] Tsai SC, Lin SJ, Chen PW, Luo WY, Yeh TH, Wang HW, et al. EBV Zta protein induces the expression of interleukin-13, promoting the proliferation of EBV-infected B cells and lymphoblastoid cell lines. Blood. 2009;**114**:109-118

[62] Hong GK, Gulley ML, Feng WH, Delecluse HJ, Holley-Guthrie E, Kenney SC. Epstein-Barr virus

lytic infection contributes to lymphoproliferative disease in a SCID mouse model. Journal of Virology. 2005;**79**:13993-14003

[63] Cannon MJ, Pisa P, Fox RI, Cooper NR. Epstein-Barr virus induces aggressive lymphoproliferative disorders of human B cell origin in SCID/hu chimeric mice. The Journal of Clinical Investigation. 1990;**85**:1333-1337

[64] Mosier DE, Gulizia RJ, Baird SM, Wilson DB. Transfer of a functional human immune system to mice with severe combined immunodeficiency. Nature. 1988;**335**:256-259

[65] Rowe M, Young LS, Crocker J, Stokes H, Henderson S, Rickinson AB. Epstein-Barr virus (EBV)-associated lymphoproliferative disease in the SCID mouse model: Implications for the pathogenesis of EBV-positive lymphomas in man. The Journal of Experimental Medicine. 1991;**173**:147-158

[66] Li H, Liu S, Hu J, Luo X, Li N, Bode AM, et al. Epstein-Barr virus lytic reactivation regulation and its pathogenic role in carcinogenesis. International Journal of Biological Sciences. 2016;**12**:1309-1318

[67] Bogedain C, Wolf H, Modrow S, Stuber G, Jilg W. Specific cytotoxic T lymphocytes recognize the immediate-early transactivator Zta of Epstein-Barr virus. Journal of Virology. 1995;**69**:4872-4879

[68] Hislop AD, Taylor GS, Sauce D, Rickinson AB. Cellular responses to viral infection in humans: Lessons from Epstein-Barr virus. Annual Review of Immunology. 2007;**25**:587-617

[69] Rist MJ, Neller MA, Burrows JM, Burrows SR. T cell epitope clustering in the highly immunogenic BZLF1 antigen of Epstein-Barr virus. Journal of Virology. 2015;**89**:703-712

[70] Scherrenburg J, Piriou ER, Nanlohy NM, van Baarle D. Detailed analysis of Epstein-Barr virus-specific CD4+ and CD8+ T cell responses during infectious mononucleosis. Clinical and Experimental Immunology. 2008;**153**:231-239

[71] Precopio ML, Sullivan JL, Willard C, Somasundaran M, Luzuriaga K. Differential kinetics and specificity of EBV-specific CD4+ and CD8+ T cells during primary infection. Journal of Immunology. 2003;**170**:2590-2598

[72] Pudney VA, Leese AM, Rickinson AB, Hislop AD. CD8+ immunodominance among Epstein-Barr virus lytic cycle antigens directly reflects the efficiency of antigen presentation in lytically infected cells. The Journal of Experimental Medicine. 2005;**201**:349-360

[73] Drouet E, Chapuis-Cellier C, Bosshard S, Verniol C, Niveleau A, Touraine JL, et al. Oligo-monoclonal immunoglobulins frequently develop during concurrent cytomegalovirus (CMV) and Epstein-Barr virus (EBV) infections in patients after renal transplantation. Clinical and Experimental Immunology. 1999;**118**:465-472

[74] Niveleau A, Bruno C, Drouet E, Brebant R, Sergeant A, Troalen F. Grafting peptides onto polystyrene microplates for ELISA. Journal of Immunological Methods. 1995;**182**:227-234

[75] Drouet E, Chapuis-Cellier C, Garnier JL, Touraine JL. Early detection of EBV infection and meaning in transplant patients. In: Touraine JL, editor. Cancer in Transplantation: Prevention and Treatment. Lyon: Kluwer Academic Publishers; 1996. pp. 201-207

[76] Sinclair AJ. bZIP proteins of human gammaherpesviruses. The Journal of General Virology. 2003;**84**:1941-1949

[77] Flemington EK. Herpesvirus lytic replication and the cell cycle: Arresting new developments. Journal of Virology. 2001;**75**:4475-4481

[78] Cayrol C, Flemington E. G0/G1 growth arrest mediated by a region encompassing the basic leucine zipper (bZIP) domain of the Epstein-Barr virus transactivator Zta. The Journal of Biological Chemistry. 1996;**271**:31799-31802

[79] Cayrol C, Flemington EK. The Epstein-Barr virus bZIP transcription factor Zta causes G0/G1 cell cycle arrest through induction of cyclin-dependent kinase inhibitors. The EMBO Journal. 1996;**15**:2748-2759

[80] Rodriguez A, Jung EJ, Yin Q, Cayrol C, Flemington EK. Role of c-myc regulation in Zta-mediated induction of the cyclin-dependent kinase inhibitors p21 and p27 and cell growth arrest. Virology. 2001;**284**:159-169

[81] Petosa C, Morand P, Baudin F, Moulin M, Artero JB, Muller CW. Structural basis of lytic cycle activation by the Epstein-Barr virus ZEBRA protein. Molecular Cell. 2006;**21**:565-572

[82] Mahot S, Fender P, Vives RR, Caron C, Perrissin M, Gruffat H, et al. Cellular uptake of the EBV transcription factor EB1/Zta. Virus Research. 2005;**110**:187-193

[83] RotheR,LiguoriL,Villegas-MendezA, Marques B, Grunwald D, Drouet E, et al. Characterization of the cell-penetrating properties of the Epstein-Barr virus ZEBRA trans-activator. The Journal of Biological Chemistry. 2010;**285**:20224-20233

[84] Habib M, Buisson M, Lupo J, Agbalika F, Socie G, Germi R, et al. Lytic EBV infection investigated by detection of soluble Epstein-Barr virus ZEBRA in the serum of patients with PTLD. Scientific Reports. 2017;**7**:10479

[85] Ensoli B, Barillari G, Salahuddin SZ, Gallo RC, Wong-Staal F. Tat protein of HIV-1 stimulates growth of cells derived from Kaposi's sarcoma lesions of AIDS patients. Nature. 1990;**345**:84-86

[86] Amorim JH, Alves RP, Boscardin SB, Ferreira LC. The dengue virus non-structural 1 protein: Risks and benefits. Virus Research. 2014;**181**:53-60

[87] Tosato G, Teruya-Feldstein J, Setsuda J, Pike SE, Jones KD, Jaffe ES. Post-transplant lymphoproliferative disease (PTLD): Lymphokine production and PTLD. Springer Seminars in Immunopathology. 1998;**20**:405-423

[88] Garnier JL, Cooper NR, Cannon MJ. Low expression of CD20 and CD23 in Epstein-Barr virus-induced B cell tumors in SCID/hu mice. The American Journal of Pathology. 1993;**142**:353-358

[89] Garnier JL, Blanc-Brunat N, Laurin D, Vivier G, Touraine JL, Sergeant A, et al. Lymphomas in the Scid/Hu mice: Effect of EBV on human B-cells in vivo. In: Touraine JL, editor. Cancer in Transplantation: Prevention and Treatment. Lyon: Kluwer Academic Publishers; 1996. pp. 209-213

[90] Montone KT, Hodinka RL, Salhany KE, Lavi E, Rostami A, Tomaszewski JE. Identification of Epstein-Barr virus lytic activity in post-transplantation lymphoproliferative disease. Modern Pathology. 1996;**9**:621-630

[91] Montone KT, Litzky LA, Wurster A, Kaiser L, Bavaria J, Kotloff R, et al. Analysis of Epstein-Barr virus-associated posttransplantation lymphoproliferative disorder after lung transplantation. Surgery. 1996;**119**:544-551

[92] Rea D, Delecluse HJ, Hamilton-Dutoit SJ, Marelle L, Joab I, Edelman L, et al. Epstein-Barr virus latent and

replicative gene expression in post-transplant lymphoproliferative disorders and AIDS-related non-Hodgkin's lymphomas. French Study Group of Pathology for HIV-associated Tumors. Annals of Oncology. 1994;**5**(Suppl 1): 113-116

[93] Rea D, Fourcade C, Leblond V, Rowe M, Joab I, Edelman L, et al. Patterns of Epstein-Barr virus latent and replicative gene expression in Epstein-Barr virus B cell lymphoproliferative disorders after organ transplantation. Transplantation. 1994;**58**:317-324

[94] Fink SE, Gandhi MK, Nourse JP, Keane C, Jones K, Crooks P, et al. A comprehensive analysis of the cellular and EBV-specific microRNAome in primary CNS PTLD identifies different patterns among EBV-associated tumors. American Journal of Transplantation. 2014;**14**:2577-2587

[95] Cohen M, Vistarop AG, Huaman F, Narbaitz M, Metrebian F, De Matteo E, et al. Epstein-Barr virus lytic cycle involvement in diffuse large B cell lymphoma. Hematological Oncology. 2018;**36**:98-103

[96] Vajro P, Lucariello S, Migliaro F, Sokal E, Gridelli B, Vegnente A, et al. Predictive value of Epstein-Barr virus genome copy number and BZLF1 expression in blood lymphocytes of transplant recipients at risk for lymphoproliferative disease. The Journal of Infectious Diseases. 2000;**181**:2050-2054

[97] Imbert-Marcille BM, Coste-Burel M, RobillardN,Foucaud-GamenJ,BillaudelS, DrouetE.Sequentialuseofparaformaldehyde and methanol as optimal conditions for the direct quantification of ZEBRA and rta antigens by flow cytometry. Clinical and Diagnostic Laboratory Immunology. 2000;7:206-211

[98] Babcock GJ, Decker LL, Freeman RB, Thorley-Lawson DA. Epstein-barr virus-infected resting memory B cells, not proliferating lymphoblasts, accumulate in the peripheral blood of immunosuppressed patients. The Journal of Experimental Medicine. 1999;**190**:567-576

[99] Miyashita EM, Yang B, Lam KM, Crawford DH, Thorley-Lawson DA. A novel form of Epstein-Barr virus latency in normal B cells in vivo. Cell. 1995;**80**:593-601

[100] Drouet E, Habib M, Agbalika F. Use of at least one biomarker for the in vitro prognosis or diagnosis of lymphoproliferative episodes associated with the Epstein-Barr virus (EBV). In: Joseph-Fourier U, editor. United States Patent Application Publication. France: Université Joseph-Fourier; 2014. pp. 1-17

[101] Oertel S, Trappe RU, Zeidler K, Babel N, Reinke P, Hummel M, et al. Epstein-Barr viral load in whole blood of adults with posttransplant lymphoproliferative disorder after solid organ transplantation does not correlate with clinical course. Annals of Hematology. 2006;**85**:478-484

[102] Wareham NE, Mocroft A, Sengelov H, Da Cunha-Bang C, Gustafsson F, Heilmann C, et al. The value of EBV DNA in early detection of post-transplant lymphoproliferative disorders among solid organ and hematopoietic stem cell transplant recipients. Journal of Cancer Research and Clinical Oncology. 2018;**144**:1569-1580

[103] Dharnidharka VR. Peripheral blood Epstein-Barr viral nucleic acid surveillance as a marker for posttransplant Cancer risk. American Journal of Transplantation. 2017;**17**:611-616

[104] Bilger A, Plowshay J, Ma S, Nawandar D, Barlow EA, Romero-Masters JC, et al. Leflunomide/teriflunomide inhibit Epstein-Barr virus

(EBV)- induced lymphoproliferative disease and lytic viral replication. Oncotarget. 2017;**8**:44266-44280

[105] Kawada JI, Ando S, Torii Y, Watanabe T, Sato Y, Ito Y, et al. Antitumor effects of duvelisib on Epstein-Barr virus-associated lymphoma cells. Cancer Medicine. 2018;**7**:1275-1284

[106] Daskalogianni C, Pyndiah S, Apcher S, Mazars A, Manoury B, Ammari N, et al. Epstein-Barr virus-encoded EBNA1 and ZEBRA: Targets for therapeutic strategies against EBV-carrying cancers. The Journal of Pathology. 2015;**235**:334-341

[107] Hartlage AS, Liu T, Patton JT, Garman SL, Zhang X, Kurt H, et al. The Epstein-Barr virus lytic protein BZLF1 as a candidate target antigen for vaccine development. Cancer Immunology Research. 2015;**3**:787-794

[108] Adhikary D, Behrends U, Boerschmann H, Pfunder A, Burdach S, Moosmann A, et al. Immunodominance of lytic cycle antigens in Epstein-Barr virus-specific CD4+ T cell preparations for therapy. PLoS One. 2007;**2**:e583

[109] Turrini R, Merlo A, Martorelli D, Fae DA, Sommaggio R, Montagner IM, et al. A BARF1-specific mAb as a new immunotherapeutic tool for the management of EBV-related tumors. Oncoimmunology. 2017;**6**:e1304338

[110] Houali K, Wang X, Shimizu Y, Djennaoui D, Nicholls J, Fiorini S, et al. A new diagnostic marker for secreted Epstein-Barr virus encoded LMP1 and BARF1 oncoproteins in the serum and saliva of patients with nasopharyngeal carcinoma. Clinical Cancer Research. 2007;**13**:4993-5000

[111] Hoebe EK, Hutajulu SH, van Beek J, Stevens SJ, Paramita DK, Greijer AE, et al. Purified hexameric Epstein-Barr virus-encoded BARF1 protein for measuring anti-BARF1 antibody responses in nasopharyngeal carcinoma patients. Clinical and Vaccine Immunology. 2011;**18**:298-304

[112] Sbih-Lammali F, Berger F, Busson P, Ooka T. Expression of the DNase encoded by the BGLF5 gene of Epstein-Barr virus in nasopharyngeal carcinoma epithelial cells. Virology. 1996;**222**:64-74

[113] Stolzenberg MC, Debouze S, Ng M, Sham J, Choy D, Bouguermouh A, et al. Purified recombinant EBV desoxyribonuclease in serological diagnosis of nasopharyngeal carcinoma. International Journal of Cancer. 1996;**66**:337-341

[114] Blay JY, Voorzanger N, Favrot M, Burdin N, Rousset F, Banchereau J. Presence of Epstein-Barr virus viral interleukin-10 in the serum of patients with non-human-immunodeficiency-virus-related diffuse large-cell non-Hodgkin's lymphomas. Blood. 1995;**86**:4702-4704

[115] Joab I, Nicolas JC, Schwaab G, de-The G, Clausse B, Perricaudet M, et al. Detection of anti-Epstein-Barr-virus transactivator (ZEBRA) antibodies in sera from patients with nasopharyngeal carcinoma. International Journal of Cancer. 1991;**48**:647-649

[116] Joab I, Triki H, de Saint MJ, Perricaudet M, Nicolas JC. Detection of anti-Epstein-Barr virus trans-activator (ZEBRA) antibodies in sera from patients with human immunodeficiency virus. The Journal of Infectious Diseases. 1991;**163**:53-56